IMISCOE Research Series

This series is the official book series of IMISCOE, the largest network of excellence on migration and diversity in the world. It comprises publications which present empirical and theoretical research on different aspects of international migration. The authors are all specialists, and the publications a rich source of information for researchers and others involved in international migration studies.

The series is published under the editorial supervision of the IMISCOE Editorial Committee which includes leading scholars from all over Europe. The series, which contains more than eighty titles already, is internationally peer reviewed which ensures that the book published in this series continue to present excellent academic standards and scholarly quality. Most of the books are available open access.

For information on how to submit a book proposal, please visit: http://www. imiscoe.org/publications/how-to-submit-a-book-proposal.

More information about this series at http://www.springer.com/series/13502

Michele Nori • Domenica Farinella

Migration, Agriculture and Rural Development

IMISCOE Short Reader

Springer Open

Michele Nori
Robert Schuman Centre for Advanced Studies
European University Institute
Firenze, Firenze, Italy

Domenica Farinella
Department of Political Sciences and Law
University of Messina
Messina, Messina, Italy

ISSN 2364-4087   ISSN 2364-4095   (electronic)
IMISCOE Research Series
ISBN 978-3-030-42862-4   ISBN 978-3-030-42863-1   (eBook)
https://doi.org/10.1007/978-3-030-42863-1

This Springer imprint is published by the registered company Springer Nature Switzerland AG.
The registered company address is: Gewerbestrasse 11, 6330 Cham, Switzerland

Acknowledgments

This book is an outcome, amongst others, of the ERC-funded project, PASTRES (www.pastres.org), which looks at how uncertainty is responded to in the pastoral world. In this book, our common interest and inspiration centres on pastoral mobilities and human migration.

Migration in agro-pastoral areas in Europe is the main focus of the book. Through the in-depth experience of the authors, the interfaces. The interfaces between agriculture, rural development, and migration studies in agro-pastoral settings have been extensively analysed across the book, with a view to understanding the reconfiguration of the human, political, and natural landscapes in the Mediterranean and beyond. These are the domains in which the authors have in-depth experience.

Amongst the many inspiring friends and colleagues, we would like to specifically thank Anna Triandafyllidou and Ian Scoones for this opportunity to reflect and report on our experiences in agro-pastoral settings.

Pastoralism, Uncertainty, Resilience

European Research Council
Established by the European Commission

We would also like to thank the many colleagues who have contributed to our thinking through stimulating discussions, ideas, suggestions, and opportunities for sharing research experiences. This long list includes Alessandra Corrado, Francesco Saverio Caruso, Carole Counihan, Nick Dines, Martina Lo Cascio, Daniela Luisi, Benedetto Meloni, Letizia Palumbo, Domenico Perrotta, Giovanni Pietrangeli,

Athanasios Ragkos, Geraldine Renaudiere, Antonio Miguel Solana, Philippe Fargues, and Ricard Zapata.

A special thanks to Ann Kingsolver, Fabio Mostaccio, Linda Pappagallo, Diane Shugart, and all the reviewers for reading and commenting on a draft of this book. Finally, an especially grateful thanks to our patient families.

Contents

Acronyms

AFN	Alternative Food Networks
CAP	Common Agricultural Policy
CIHEAM	Centre International de Hautes Etudes Agronomiques Méditerranéennes
CSIC	Consejo Superior de Investigaciónes Cientificas
ECHR	European Court of Human Rights
ELSTAT	Hellenic Statistical Authority
ERC	European Research Council
ETT	Empresas de Trabajo Temporal
EU	European Union
EUI	European University Institute
EUMed	Mediterranean European Union
FAO	Food and Agriculture Organization
FNAAC	Framework National Agreement for Agriculture Campaigns
HNV	High Nature Value
GATS	General Agreement on Trade and Services
GVP	Gross Value of Production
IDS	Institute of Development Studies, University of Sussex
INE	Instituto Nacional de Estadistica
INEA	Istituto Nazionale di Economia Agraria
INPS	Istituto Nazionale di Previdenza Sociale
IPCC	Intergovernmental Panel on Climate Change
ISTAT	Istituto Italiano di Statistica
MENA	Middle East and Northern Africa
MPC	Migration Policy Centre
OMT	Observatorio del Mercado de Trabajo
OPI	Observatorio Permanente de la Inmigración
SMEs	Small and Medium Enterprises
OECD	Organisation for Economic Co-operation and Development
OPR	Osservatorio Placido Rizzotto
UNEP	United Nations Environment Programme

SFSC	Short Food Supply Chains
UAA	Utilized Agricultural Area
UfM	Union for the Mediterranean
WTO	World Trade Organization

Chapter 1
Rural World, Migration, and Agriculture in Mediterranean EU: An Introduction

This book investigates the dynamics that are reshaping human and natural landscapes in the European agrarian world, with a specific focus on Mediterranean Europe. We focus here on more marginal rural settings, where the potential for agricultural intensification is structurally limited. These areas in particular have suffered from the geographical and socio-economic polarization of development patterns and have paid a relevant burden to the recent crisis.

In these areas, immigration has, to an extent, helped counterbalance the dynamics of an ageing and declining local population, with immigrant communities today relevant not only as an agricultural workforce, but also as new citizens of rural communities.

Contemporary migrations from and to rural areas are to be analysed in relation to the incorporation of agrarian systems into global markets, agricultural governance, and local territories' struggle between innovation and resilience.

Disentangling the critical relationships between the conditions of agricultural work, rural development paradigms, labour markets, and migration policies represents a necessary step to understand the ongoing dynamics of rural mobility and to suggest opportunities and solutions that might accommodate the different interests and needs in Mediterranean societies. The interface between agriculture and migration is fertile, not only in academic terms, but also in socio-economic and political ones.

M. Nori, D. Farinella, *Migration, Agriculture and Rural Development*, IMISCOE
Research Series, https://doi.org/10.1007/978-3-030-42863-1_1

1.1 Scope and Aims of the Book[1]

This book aims to introduce students and practitioners to migration from the perspective of agriculture and rural development. Intense territorial polarizations in recent decades and the resulting reconfiguration of the agrarian world have resulted in emigration increasingly representing a key livelihood strategy for rural households. Today, across the globe, rural youth seek better conditions and opportunities often away from their communities of origin. The implications for rural development are diverse and controversial, in social, economic, as well as environmental terms.

The Mediterranean represents an appropriate setting for exploring the interfaces between migratory flows, agriculture, and rural development in a wider perspective as the region is simultaneously an area of emigration, immigration, and transit for migrants. Migration is not new here, as the Mediterranean is historically at the crossroads of three continents, with mobility characterized by different triggers, pace, and trajectories in diverse periods.

This book focuses on the specificities of the agrarian world in the Mediterranean EU, which is increasingly populated by immigrants who originate from poorer southern and eastern regions and who have come to live and work in the European countryside. Its chapters analyse the role of migratory flows in tackling the social and economic mismatch of rural labour markets in critical societal domains such as the production of food and the management of natural resources.

In more marginal rural settings, where the potential for agricultural intensification is structurally limited and the consequences from recent economic shocks are higher, immigrant communities play a particularly important role in filling the socio-economic gaps left by the national population.

These dynamics seem to reproduce patterns of mobility typical of the agrarian world and represent an invaluable opportunity for territories and sectors that would otherwise face abandonment and desertification. The significance of immigrant communities notwithstanding, there are problems with their recognition and integration as both agricultural workers and as rural citizens.

After offering a broad overview of the restructuring patterns that have affected agriculture and rural areas in the EU, this book analyses contemporary rural

---

[1]This book is the result of joint and equal contributions by the authors.

In Chap. 1, Sects. 1.1 and 1.2 should be attributed to Michele Nori. Sections 1.3 and 1.4 to Domenica Farinella.

In Chap. 2, Sects 2.1, 2.2 and 2.3 should be attributed to Domenica Farinella. Sections 2.4 and 2.5 to both authors.

In Chap. 3, Sects 3.1, 3.6.1, 3.6.2 and Appendix 1 should be attributed to Michele Nori. Sections 3.3, 3.5, 3.6.3 and 3.7 should be attributed to Domenica Farinella. Sections 3.2 and 3.4 to both authors.

In Chap. 4, Sect. 4.1 should be attributed to Michele Nori. Sections 4.3 and 4.4 to Domenica Farinella. Section 4.2 and 4.5 to both authors.

Chap. 5 should be attributed to Michele Nori.

Both authors contributed equally to Chap. 6 and Conclusions.

migrations and the emergence of immigrants in relation to the incorporation of agrarian systems into global markets and European agricultural governance.

Set between tradition, innovation, and resilience, rural areas express in fact the contemporary contradictions of the neoliberal global world. On the one hand, these are the site of exodus, population decline, economic crisis, and land abandonment or overexploitation. On the other, these represent the space for local movements for autonomy, peasant agriculture, and rural revitalization.

While most of the existing literature focuses on the role of immigrant workers in intensive agricultural production, little attention has been given to agriculture systems in more marginal and remote settings: the mountainous territories inner regions and the islands that comprise a large part of the Mediterranean region. Most of the examples and cases reported in this volume refer to the specificities of agro-pastoral systems in the EUMed (Greece, Spain, and Italy), the domain and region which the two authors have researched intensively in recent years. The reconfiguration of agriculture and rural landscapes will be assessed through the lens of agro-pastoralism; the in-depth exploration of the dynamics surrounding immigrant shepherds will shed light on contemporary phenomena reshaping the agrarian world.

Agro-pastoralism—that is, the extensive rearing of livestock complemented by farming—represents a primary production system in marginal territories and thus the main source of income, employment, and livelihood in areas where more intense and capital-based agriculture is not feasible. In these regions agro-pastoral practices are critical not only for productive purposes, but also to manage landscape and ecological resources and preserve local knowledge of the environment. These are captured by the social and environmental services and benefits associated with agro-pastoral systems and practices.

Despite growing societal appreciation, agro-pastoralism is becoming less and less attractive, especially for young people; shepherding is very demanding, while earnings are meagre. The limited services and facilities available in agro-pastoral territories represent further disincentives for local youth to engage in such activity. In these contexts, immigrants have come to provide skilled labour at relatively low costs. Without foreign workers, many agro-pastoral farms would face great difficulty in pursuing their activities. Evidence shows that immigrants not only participate in productive activities, but they also represent an overall strategic resource for the social and economic development of marginal areas as well as for reproducing local societies.

However, this solution is temporary. Harsh living and working conditions, illegality, and socio-economic vulnerability represent important disincentives for immigrants to remain. Moreover political, economic, and administrative problems provide major constraints and obstacles to the broader integration of rural immigrants. Geographical mobility does not evolve into social mobility; workers face difficulties in scaling up in social and economic terms, with little prospect of eventually graduating as farmers, entrepreneurs, citizens. This raises concerns about the sustainability of the current dynamics as both agricultural farms and rural communities in the EU are facing serious problems of generational renewal and related syndromes of abandonment and socio-economic decline despite the consistent "rural welfare" system put in place through the Common Agricultural Policy (CAP).

In the chapters that follow, we undertake a traditional origin-destination perspective to investigate the implications and impacts of rural migratory flows in the different settings. We are aware that more recent and up-to-date approaches on migration studies undertake more fluid and multi-sited perspectives whereby migrants contemporarily inhabit different realities and migration is just part of wider mobility processes. Migrants' agency is nowadays central to how academics theorize migration, how government officials design policies, and how activists devise campaigns to influence policies. However, rather than just focusing on migrants themselves, our focus is on the dynamics and the processes affecting agrarian societies and the rural world.

1.2   The Plan of the Book

The book is structured to guide readers through different themes aimed at offering a comprehensive, consistent, and concise set of elements informing the scientific and policy debates on agriculture and rural migrants. Reference to more consistent and elaborated perspectives and analyses is made for the different domains and realities throughout the text.

In the next chapter we analyse the three processes characterizing agriculture and rural areas from the 1950s to the present: (i) agricultural modernization and polarization; (ii) the restructuring of agri-food chains in the global market; and, (iii) the institutionalization of the agrarian world, including the role of the Common Agricultural Policy (CAP). Our focus is on EUMed countries (Greece, Spain, and Italy), which present some specific and characterising features.

The third chapter focuses on the ambivalent nature of contemporary agricultural migration in European rural areas. The growing presence of immigrants in these areas is a direct result of the restructuring of agriculture and global agri-food chains. Evidence indicates that while agricultural work and rural settings are decreasingly attractive to local populations, they represent a favourable environment to international newcomers as they offer a better chance to access livelihood resources compared to urban areas. The chapter starts by providing a basic understanding of the growing centrality of immigrant workers in agriculture and as citizens of rural communities. The specificities of the Mediterranean migration model are then assessed, followed by a more in-depth analysis of the agricultural sector and the broader rural world in Greece, Spain, and Italy.

In Chap. 4 we provide a framework for assessing and analysing ongoing rural migration dynamics from the perspective of areas of destination, with a view to answering to the following questions: What happens to rural areas of destination? What are the impacts on the local economy and society? Which are the practices, programs, and policies that underpin the presence and integration of migration? What is recent experience revealing on these matters?

In particular, we focus on the more marginal, isolated, remote areas of the EUMed where the contributions of immigrants are critical for the sustainability and repro-duction of local societies. In these areas, immigrant communities could, in fact, represent a strategic asset with an eye to offset processes of rural population decline and abandonment of the agrarian world. The chapter progresses through several cases and experiences related to processes and practices of inclusion and integration of immigrants in Italian settings.

Chapter 5 looks at the implications, impacts, and consequences of rural migration on the areas of origin, where oftentimes portions of the family—and some of its assets—remain. Why do people leave their rural communities? What are the drivers and triggers inducing such emigration? What are the implications for those remaining behind? What are the impacts on local communities and development patterns?

In Chap. 6 we present the specific case of immigrant shepherds in agro-pastoral areas of Greece, Spain, and Italy with a view to unfold and assess the contributions of immigrant communities to the sustainable development of marginal territories.

Final chapter concludes that agriculture and the rural world represent a relevant setting to tackle the challenges linked to the intense migration that is reshaping our societies. Evidence shows that particularly in rural settings, immigrants play a key role in maintaining and reproducing local societies and their embedded heritage. Indeed, the agriculture sector and the rural world hold important potentials for fostering the economic and social integration of migrants and refugees, as attested by the several programs and initiatives we explore.

1.3 Methodology

This volume aims at helping students and researchers to evolve the scientific debate around the theme of migrants in rural areas by identifying the main research areas and most important works carried out so far in this domain. It is the result of a research path on which the two authors started in 2014; it combines a review of existing literature with empirical data based on original field research using both quantitative and qualitative methods.

The literature review has been instrumental to unfold the debate about rural migrations and frame the main theoretical questions. Concrete field experiences have been used to support a more empirical understanding of the themes brought to discussion. The use of the "text box" throughout the book helps highlight specific aspects in order to clarify concepts and approaches or to present emblematic empirical cases.

The quantitative and diachronic analysis of data is used to describe the general framework and the main trends. Data used have been mainly sourced from EuroStat, Caritas, Oxfam, the Hellenic Statistical Office (ELSTAT), the Greek Ministry of Migration Policy, the Hellenic Foundation for European and Foreign Policy (ELIAMEP), the Instituto Nacional Estadistico (INE), the Observatorio Mercado del Trabajo (OMT), the Observatorio Permanente de la Inmigración (OPI), the Istituto Nazionale di Statistica (ISTAT), the Osservatorio Placido Rizzotto (OPR), the Istituto Nazionale Economia Agraria (INEA) and the Istituto di Servizi per il Mercato Agricolo Alimentare (ISMEA).

The book also presents the results of extensive fieldwork based on qualitative methods such as ethnographic observation, in-depth interviews, and semi-structured questionnaires. These qualitative data have been elaborated through several projects undertaken by the authors on the relations between rural areas and immigration. Main experiences include:

- the EU project Food Track (VP/2016/004) "A transparent and traceable food supply chain for the benefit of workers, enterprises and consumers: the role of a multi-sectoral approach of industrial relations and corporate social responsibility." Funded by European Commission – DG Employment, Social Affairs & Inclusion. International Coordinator: FILCAMS-CGIL. D. Farinella (University of Cagliari) was coordinator for Italian Research Unit.
- the Open Society European Policy Institute project on immigrants' exploitation in Italian agriculture which investigated the restructuring of agri-food chains and the factors pushing farmers to recruit migrant workers irregularly and thus profiting from their vulnerable conditions. M. Nori collaborated as a research assistant in 2018. The full report of the project is referred to as: Corrado, A., Palumbo L., Caruso F. S., Lo Cascio M., Nori M., Triandafyllidou, A., (2018). *Is Italian Agriculture a 'Pull Factor' for Irregular Migration – and, if so, why?* Open Society Foundations. It can be accessed at: https://www.opensocietyfoundations. org/sites/default/files/is-italian-agriculture-a-pull-factor-for-irregular-migration-20181205.pdf
- the Forum on Agriculture, Rural Development and Migration in the Mediterranean, an interagency initiative jointly organised by EUI, FAO, CIHEAM, and UfM to discuss migrations in the Mediterranean from the perspective of rural and agricultural development. The initiative aimed to provide policy recommendations and establish a regional multi-stakeholder platform for decision-makers at different levels. M. Nori was amongst the initiative coordinators in 2017–18. The full report of the initiative is available at http://cadmus.eui.eu/bitstream/handle/1814/60473/GGP_RR_2019_01.pdf?sequence=1&isAllowed=y
- the Strategia Nazionale Aree Interne of the Italian Government (http://old2018. agenziacoesione.gov.it/it/arint/), for which M. Nori is an external collaborator.

The data on the pull and push factors for rural emigrations and on the implications in the communities of migrants' origin refer to:

- the Rural Youth Migration initiative funded by Italian Development Cooperation and FAO and implemented by the EUI Migration Policy Centre. The project aimed to enhance the understanding of rural youth emigration in Tunisia, with a view to facilitating positive impacts on food security, agriculture, and development in rural areas. M. Nori collaborated as a research assistant on the qualitative aspects of the research in 2017 through an innovative, participatory method that combined quantitative and qualitative components. The project report is referenced as Zuccotti C.V., Geddes A.P., Bacchi A., Nori M., Stojanov R., 2018. *Rural Migration in Tunisia. Drivers and patterns of rural youth migration and its impact on food security and rural livelihoods in Tunisia.* Food and Agriculture Organization of the United Nations. Rome. It can be accessed at: http://www.fao.org/rural-employment/work-areas/migration/rym-project/en/.

The section on agro-pastoralism is inspired by:

- the TRAMed – Transhumances in the Mediterranean project, funded by the EU Marie Curie program (contract ES706/2014). The research was concerned with assessing ongoing dynamics affecting pastoralism in the Mediterranean in order to provide a more effective understanding of the presence and contribution of immigrants in this domain, with the view to contribute to the development of appropriate policies at the local and European levels. M. Nori was the project coordinator and principal researcher during the period 2015–2017. D. Farinella was an external collaborator and consultant for the data analysis. During this project, two different sets of semi-structured interviews have been collected with closed and open questions addressed to stockowners (110) and foreign workers (35) in parts of Greece (Peloponnesus and Macedonia), Spain (Cataluña), Italy (Piedmont, Triveneto and Abruzzo), and Provence in France.
- the National Project "Changes of Sardinian pastoralism: shepherds and Romanian workers," funded by the Autonomous Region of Sardinia and the University of Cagliari from 2015 to 2018. D. Farinella was the project coordinator and principal researcher. This project was based on ethnographic observation methods and in-depth interviews (using the technique of collecting life stories) of Sardinian breeders (more than 100) and foreign workers (21).
- The PASTRES project (www.pastres.org), funded by the EU European Research Council, jointly implemented by EUI and IDS of the University of Sussex. M. Nori is a research associate within the project and D. Farinella is an affiliate researcher. Drawing insights from across continents, the project is asking what lessons can we learn for global challenges from pastoral systems responding to uncertainty? In six pastoral regions of the world, the project explores responses to uncertainty in three domains: environment/resources, markets/commodities, and institutions/governance. The challenge is to draw out wider lessons to inform knowledge and decision-making in other societal dimensions where uncertainty is central, including climate and environmental change, financial and commodity markets, response, critical infrastructures, migration policy, and security and conflict.

1.4   The Main Theoretical Issues

Many of the processes and changes described in the book refer to different theoretical and analytical approaches concerning particular issues such as agriculture, rural areas, labour markets, and migration. Without any ambition of completeness, in this section we try to provide the reader with a general framework of these theoretical issues organized by topics.

1.4.1   Agriculture and the Rural Space

For decades, the dominant paradigm in agricultural studies and national policies has been agricultural modernization. In the broad context of modernization theory, this paradigm is founded on the idea of development and civilization as a linear and positive transition from "pre-modern," "traditional," and "rural" society to "modern," "industrial," and "urban" (Martinelli 2005; He 2012). Many criticisms can be levelled at this approach, including naive positivism and ethnocentrism (considering Western countries as the *one best way* of "modernization") (Escobar 1995). The methodological approach combines structural functionalism with the rational actor model and methodological individualism. The research schemes are based on a wide variety of methods, including quantitative methods, testing cause-effect relations, and multivariate analysis.

By the end of the second world war, agricultural modernization theories (see He 2012: 509–528) promoted the agricultural efficiency as a strategic issue for the economic development of the national states. Agricultural modernization implied the application of scientific principles and technological innovations to agricultural activities, making them rational, replicable, predictable, and efficient. The farmer was supposed to become a rational entrepreneur producing food for the market. The agricultural modernization approach suggested measures as specialization and monoculture, standardization and mechanization, intensification, large-scale and mass production, electrification, irrigation, use of technology, and chemicals and fertilizers. From the 1960s onwards, a specific set of studies introduced the term "green revolution" indicating the application of a wide range of technologies – from genetics to mechanics and chemicals – in "developing" countries to increase farm productivity.

For these theorists, agricultural innovations should have guaranteed mass and low-cost food, solving the difficult balance of food supply and demand in the national States, while bolstering their respective national interests and contributing to social stability. The decline of the percentage of agricultural workers was offset by the increase in their productivity.

Another important requirement of agricultural modernization is neoliberalism: marketization and international trade, commodification of all productive factors (land and labour) as well as of agricultural products stimulate the efficiency of local agricultural systems and lower final prices.

After the 1970s, agricultural modernization theories tried to include in the analysis the new challenges posed by the information and biotechnology revolution, the emergence of globalization, and environmental issues. The growing ecological constraints associated with the non-reversible consumption of environmental resources, climate change, pollution, agricultural squeeze, and new exploitation of workers required the adaptation of "development" within "sustainability." Following this path, the new agricultural modernization approach theorizes a sustainable and knowledge-based agriculture, oriented to informatization, renewable energy sources, biotechnologies, and the high-tech revolution. The new studies analyse the ecologization and the transition to green and organic practices, based on diversification and plant-type agriculture (see Christoff 1996; Cohen 1997; Mol 2001).

In contrast with the agricultural modernization approach, in the mid-Seventies, new agri-food studies were stimulated by Marxist tradition and world-system analysis (Wallerstein 1979; Hopkins and Wallerstein 1994). Van der Ploeg's approach of "new peasantries" (2008, 2013) considers the "peasant mode of farming" distinctive with respect to entrepreneurial and capitalistic agriculture. "Peasant farming" is a small-scale, familiar, and intensive-labour based model. The peasant model is part of the global market but is in constant struggle for autonomy and to resist commodification, enhancing its internal and redundant elements such as self-production and self-consumption for inputs and outputs, embeddedness in the local environment, coproduction, reciprocity, and non-monetary exchanges, multifunctionality and pluriactivity, and family labour. Following Van der Ploeg (2008:1):

> Peasant agriculture "is basically built upon the sustained use of ecological capital and oriented towards defending and improving peasant livelihoods. Multifunctionality is often a major feature. Labour is basically provided by the family (or mobilized within the rural community through relations of reciprocity), and land and the other major means of production are family owned. Production is oriented towards the market as well as towards the reproduction of the farm unit and the family.

The importance of the peasant model is emphasized as a subsistence and autonomy strategy in marginal rural areas, thus highlighting how agro-pastoralism is peasant farming. Other approaches are inspired by the global commodity chains (Gereffi and Korzeniewicz 1990, 1994) and the food regime analysis (Friedmann and McMichael 1989; McMichael 2013), focusing on the role of agriculture in the development of the capitalist world economy and emphasizing the phenomena of accumulation by dispossession related to food relationships in international trade. The incorporation of agriculture into the global supply chains has strengthened the power of corporations and large-scale retail trade in which private forms of regulation prevail. The methodological approach is prevalently based on historical and comparative analysis. As McMichael (2009:140) synthetizes:

> the food regime concept historicized the global food system: problematizing linear representations of agricultural modernization, underlining the pivotal role of food in global political-economy and conceptualizing key historical contradictions in particular food regimes that produce crisis, transformation and transition.

For researchers in this area, the current food regime is based on the "transnational restructuring of agricultural sectors [...] through (i) intensification of agricultural specialization (for both enterprises and regions) and integration of specific crops and livestock into agro-food chains dominated at both ends by increasingly large industrial capitals; and (ii) a shift in agricultural products from final use to industrial inputs for manufactured foods" (Friedmann and McMichael 1989: 105).

The methodological approach is both structural and critical: agriculture and the farmers are the weakest part of a large chain whose components are interrelated and mutually dependent. These researchers analyse the creation and distribution of value into the agri-food supply chains (Burch and Lawrence 2007), focusing on the asymmetries and concentrations of power as well as on the agricultural squeeze.

Another approach is the Rural Development model. This paradigm was developed by OECD, EU, and other transnational institutions. It takes a pragmatic and policy-oriented approach aimed at overcoming the sectorial and productivist approach typical of the classic theories of agricultural modernization (OECD 2006, 2016; Van der Ploeg and Mardsen 2008). In this sense, Rural Development implies a "new developmental model for the agricultural sector" (Van der Ploeg et al. 2000: 392) responding to contemporary challenges, such as, in particular, the rural exodus and land abandonment, the global competition amongst territories, pollution and climate change, the restructuring of the rural economy (with the decrease in agricultural work), and the high-tech revolution.

Rural Development is a post-productivist paradigm: the basic idea is that agriculture is no longer the main source of income and labour in rural regions; the direct correspondence between agriculture and rurality is challenged here, as the latter should be analysed considering its complexity and autonomy. On one hand, rural areas are less competitive in economic and political terms compared to urban settings, especially in times of public spending cuts, but rural population demand the same services as those in urban areas. On the other hand, in a context of globalization and increasingly segmented social demands, rural areas provide services, activities, and products of high ecological and ethical value that are not available in urban areas. In this sense, rural areas have factors of attractiveness for an emerging class of consumers: the rural users.

Another fundamental aspect of the rural development paradigm is multifunctionality in agriculture (OECD 2001; Van Huylenbroeck and Durand 2003; Wilson 2007; Marsden and Sonnino 2008). This refers to the idea that agriculture has other functions in addition to food production and that it produces positive externalities, as in particular non-trade benefits and local collective goods such as ecological services (e.g. environmental protection, landscape management, food security, ecological biodiversity preservation) as well as other goods and services of high value for communities such as tourism services, training, education, and energy.

Following the OECD definition (2006), Rural Development is an integrated, multi-sectoral, place-based and multi-actor approach that aims to exploit the varied and localized potentials of rural areas, supports the empowerment of local communities, and moves from the passive logic of subsidies to the active logic of investments. On the methodological level, Rural Development is community-place-based and uses participatory research methods (Chambers 1983).

1.4.2 Labour Market and Migration Studies

To explore the dynamics between agriculture and migration in the Mediterranean region we refer to various analytical approaches. The Mediterranean area's capitalism and labour market are framed using suggestions coming from the Varieties of Capitalism (Hall and Soskice 2001; Amable 2003; Molina and Rhodes 2007), and the Welfare State regimes approaches (Esping-Andersen 1990; Ferrera 1996, 2010; Castels et al. 2010).

The Varieties of Capitalism is a theoretical perspective of New Political Economy aimed at comparing different national capitalisms by highlighting the role of economic, political, and institutional factors. The approach is actor-centred and considers the political economy as an arena "populated by multiple actors, each of whom seeks to advance its interests in rational way in strategic interaction with the others" (Hall and Soskice 2001: 6). Specifically, this approach emphasizes the key role of the firm as the agent of adjustments in different aspects of socio-economic life such as corporate governance, labour relations, technological change, and international competition.

The Welfare State regimes literature begins in 1990 with the publication of Esping-Andersen's book *The Three Worlds of Welfare Capitalism*, in discussion with the earlier contribution of Titmuss (1963). This comparative approach analyses the different ways of organization and functioning of national welfare states. Using Polanyi's theoretical tripartite division (state, market, community) (Polanyi 1944), each local welfare regime is seen as a differentiated combination of these three elements. On the methodological level, it used a comparative method based on categories built according to the Weberian "ideal-type" scheme.

A specific stream of this approach focuses on specificities of the "Latin" or "Mediterranean" model of welfare state and on its transformations (Leibfried 1992; Castles 1993; Ferrera 1996, 2010).

To analyse the Mediterranean labour market and the role of immigrants, we refer in a critical way to the Marxian theory of the Reserve Army. Marx introduced this theory as a general feature of the capitalist system. In his analysis, the "industrial reserve army" is a surplus population of unemployed and potentially available to work (therefore pressing on the workers) as an effect of capitalistic accumulation and change in the capital composition due to mechanization and productivity improvements:

> The industrial reserve army, during the periods of stagnation and average prosperity, weighs down the active labour-army; during the periods of over-production and paroxysm, it holds its pretensions in check. Relative surplus population is therefore the pivot upon which the law of demand and supply of labour works. It confines the field of action of this law within the limits absolutely convenient to the activity of exploitation and to the domination of

capital. [...] Capital works on both sides at the same time. If its accumulation, on the one hand, increases the demand for labour, it increases on the other the supply of labourers by the "setting free" of them, whilst at the same time the pressure of the unemployed compels those that are employed to furnish more labour, and therefore makes the supply of labour, to a certain extent, independent of the supply of labourers (Marx 1974: 598).

In the Marxist analysis, the Reserve Army suggests the labour market is segmented in different sub-groups of workers who are marginal and precarious to varying degrees. Using these suggestions and criticizing the neoclassical approach to the labour market based on the rational actor a group of scholars (Doeringer and Piore 1971; Reich et al. 1973; Harrison and Sum 1979; Piore 1979) in the late 1960s and early 1970s developed the theory of the dual/segmented labour market. According to this approach, there is a groove between a primary labour market (with high labour productivity)—in which the employers (often unionized) possess high degrees and are guaranteed salaries, labour rights, and stable employment—and a secondary and peripheral labour market comprised of large precarious subsectors characterized by unstable manual and unskilled work with low-productivity and low wages.[2] The latter represents the weakest social categories such as low-skilled workers, women and youth, and migrants who often are in a subaltern position in the labour market that exposes them to unemployment or under-paid jobs.

Specifically, Castles and Kosack (1973) argue that migrant workers serve as a "reserve army of labour" and Piore (1979) highlights the role of immigrants in segmented labour markets. Making use of quantitative methods and network analysis, many scholars focused on the ethnicitization of some market niches, the migratory chain, as well as the effects of displacement and replacement of migrant labour (Portes and Bach 1985; Portes and Jensen 1989; Waldinger 1994; Waldinger and Lichter 2003; Reyneri 2004; Anderson and Ruhs 2010; Ambrosini 2013; Avola 2015; Fullin 2016). These studies are often accused of considering the migrant only as worker, analysing his "functional" role in the local market.

Another theoretical approach to migration cited in this book is that of the "structural drivers." It is based on the idea there are push and pull factors influencing the migration fluxes. Introduced by Lee (1966), this approach classifies the drivers for migration, distinguishing between those that attract immigrants and those that reject them, leading them to eventually emigrate. As Saitta summarises (2008: 137):

> In short, [...] people migrate to a specific country for: a) the characteristics of the area of origin; b) the characteristics of destination area; c) for obstacles that hinder the movement; d) for the internal differentiation of the population (or the social perception of the categories of poverty and wealth). According to Lee, in this framework the analyst's task would be to identify the relevant variables to influence a rational subject to emigrate or stay in a house. Identify these variables, manage the organization within two categories with respect to the decision to leave these determine a negative (push) or positive choice (pull).

---

[2]Furthermore, Esping-Andersen (1999: 111) explains the polarization in the labour market and the entrapment of bad jobs (labour-intensive and low-wage) with "Baumol cost-disease problem" that "will come about because, in the long haul, productivity grows on average much faster in manufacturing than in (most) services."

One of the weaknesses of this neoclassical economic theory and its underlying rational-actor paradigm is its mechanic and direct association between social positions and practices, abstractly assuming that all migrants act the same way.

New perspectives in migration studies evolve from a critique to these deterministic theories, biased by a strong focus on the economic dimension (Massey et al. 1998; Arango 2000; see Kararakoulaki et al. 2018). The limitation of the push/pull approach is that it considers the migration between two countries as driven by a wage gap among geographical areas (Sjaastad 1962; Todaro 1969; Jennissen 2007). In the Marxist approach, the migration is affected by the capitalist development in the global market (Massey et al. 1993). Following the dual labour market theory, in developed countries there is always a demand for migrant workers (Piore 1979; Massey et al. 1993; Jennissen 2007). The main criticism of these theories is "they are overtly focused on why some people move whilst ignoring why others do not, as well as a lack of attention to state policies as influencers of migration. As Arango (2000) notes, migration is "both very complex and straightforward." General explanations are therefore bound to be "reductionistic" (Karakakoulaki et al. 2018: 5).

Another criticism to these theories is that it underestimates migrants' subjectivity and considers them as trapped in a substantive and abstract vision in which they are treated as objects and as quantities. These theories risk to reproduce stereotypes as they often reify migrants' behaviours, activities, and preferences as if these were permanently inscribed in a "sort of biological and cultural essence" (Accardo 2006).

Many recent critical approaches speak of migrant subjectivity in transnational mobility (Anderson 2009; Andrijasevic and Anderson 2009; Conradson and Mckay 2007; Casas-Cortes et al. 2015), using post-colonial suggestions. These studies "investigate the construction of subjectivities in relation to both oppressive and affirmative power dynamics and are working towards a theory of agency that encourages us to think in more nuanced ways about how norms and discourses are inhabited and transformed" (Andrijasevic and Anderson 2009: 366). Many research techniques are based on ethnographies and in-depth interviews.

This book is articulated within this wide theoretical framework.

References

Accardo, A. (2006). *Introduction à une sociologie critique. Lire Pierre Bourdieu*. Marseille: Agone.

Amable, B. (2003). *The diversity of modern capitalism*. Oxford: Oxford University Press.

Ambrosini, M. (2013). *Irregular migration and invisible welfare*. Basingstoke: Palgrave Macmillan.

Anderson, B. (2009). What's in a name? Immigration controls and subjectivities: The case of au pairs and domestic worker visa holders in the UK. *Subjectivity, 29*, 407–424.

Anderson, B., & Ruhs, M. (Eds.). (2010). *Who needs migrant workers? Labour shortages, immigration, and public policy*. Oxford: Oxford University Press.

Andrijasevic, R., & Anderson, B. (2009). Conflicts of mobility: Migration, labour and political subjectivities. *Subjectivity, 29*, 363–366.

Arango, J. (2000). Explaining migration: A critical view. *International Social Science Journal, 52* (165), 283–296.

Avola, M. (2015). The ethnic penalty in the Italian labour market: A comparison between the Centre-North and South. *Journal of Ethnic and Migration Studies, 42*(11), 1746–1768.

Burch, D., & Lawrence, G. (2007). *Supermarkets and agri-food supply chains transformations in the production and consumption of foods*. Cheltenham: Elgar.

Casas-Cortes, M., Cobarrubias, S., De Genova, N., Garelli, G., Grappi, G., Heller, C., Hess, S., Kasparek, B., Mezzadra, S., Neilson, B., Peano, I., Pezzani, L., Pickles, J., Rahola, F., Riedner, L., Scheel, S., & Tazzioli, M. (2015). New keywords: Migration and Borders. *Cultural Studies, 29*(1), 55–87. https://doi.org/10.1080/09502386.2014.891630.

Castels, F. G., Leibfried, S., Lewis, J., Obinger, H., & Pierson, C. (Eds.). (2010). *The Oxford handbook of welfare state*. Oxford: Oxford University Press.

Castles, F. (1993). *Families of nations. Patterns of public policy in Western democracies*. Aldershot: Hants.

Castles, S., & Kosack, G. (1973). *Immigrant workers and class structure in Western Europe*. London: Institute of Race Relations, Oxford University Press.

Chambers, R. (1983). *Rural development: Putting the last first*. Harlow: Prentice Hall.

Christoff, P. (1996). Ecological modernisation, ecological modernities. *Environmental Politics, 5* (3), 476–500.

Cohen, M. (1997). Risk society and ecological modernisation. Alternative visions for post-industrial nations. *Futures, 29*(2), 105–119.

Conradson, D., & Mckay, D. (2007). Translocal subjectivities: Mobility, connection, emotion. *Mobilities, Special Issue: Translocal Subjectivities: Mobility, Connection, Emotion, 2*(2), 167–174.

Doeringer, P., & Piore, M. (1971). *Internal labor markets and manpower analysis*. Lexington: Mass.

Escobar, A. (1995). *Encountering development: The making and unmaking of the third world*. Princeton: Princenton University Press.

Esping-Andersen, G. (1990). *The three worlds of welfare capitalism*. Princeton: Princeton University Press.

Esping-Andersen, G. (1999). *Social foundations of postindustrial economies*. New York: Oxford University Press.

Ferrera, M. (1996). The "Southern Model" of welfare in social Europe. *Journal of European Social Policy, 6*(1), 17–37.

Ferrera, M. (2010). The South European countries. In Castels et al. (Eds.), *The oxford handbook of welfare state* (pp. 616–629). Oxford: Oxford University Press.

Friedmann, H., & McMichael, P. (1989). Agriculture and the state system: The rise and decline of national agricultures, 1870 to the present. *Sociologia Ruralis, 29*(2), 93–117.

Fullin, G. (2016). Labour market outcomes of immigrants in a South European country: Do race and religion matter? *Work Employment and Society, 30*(3), 391–409.

Gereffi, G., & Korzeniewicz, M. (1990). Commodity chains and footwear exports in the semiperiphery. In W. G. Martin (Ed.), *Semiperipheral states in the world-economy* (pp. 45–68). Westport: Greenwood Press.

Gereffi, G., & Korzeniewicz, M. (Eds.). (1994). *Commodity chains and global capitalism*. Westport: Praeger.

Hall, P. A., & Soskice, D. (2001). *Varieties of capitalism: The institutional foundations of comparative advantage*. Oxford: Oxford University Press.

Harrison, B., & Sum, A. (1979). The theory of "dual" or segmented labor markets. *Journal of Economic Issues, 13*(3), 687–706.

He, C. (2012). Modernization science. In *The principles and methods of national advancement*. Heidelberg/Dordrecht/London/New York: Springer.

Hopkins, T. K., & Wallerstein, I. (1994). Commodity chains: Construct and research. In G. Gereffi & M. Korzeniewicz (Eds.), *Commodity chains and global capitalism* (pp. 17–20). Westport: Praeger.

Jennissen, R. (2007). Causality chains in the international migration systems approach. *Population Research and Policy Review, 26*, 411–436.

Karakoulaki, M., Southgate, L., & Steiner, J. (2018). *Critical perspectives on migration in the twenty-first century*. Bristol: E-International Relations.

Lee, E. (1966). A theory of migration. *Demography, 3*, 47–57.

Leibfried, S. (1992). Towards a European welfare state. In Z. Ferge, & J. E. Kolberg (Eds.), *Social policy in a changing Europe* (245–279). Boulder: Westview Press.

Marsden, T., & Sonnino, S. (2008). Rural development and the regional state: Denying multifunctional agriculture in the UK. *Journal of Rural Studies, 24*, 422–431.

Martinelli, A. (2005). *Global modernization: Rethinking the project of modernity*. London: Sage.

Marx, K. (1974 [1867]). *Capital* (Vol. I). London: Lawrence and Wishart.

Massey, D., Arango, J., Hugo, G., Kouaouci, A., Pellegrino, A., & Taylor, J. (1993). Theories of international migration: A review and appraisal. *Population and Development Review, 19*(3), 431–466. https://doi.org/10.2307/2938462.

Massey, D., Arango, J., Hugo, G., Kouaouci, A., Pellegrino, A., & Taylor, J. (1998). *Worlds in motion: Understanding international migration at the end of the millennium*. Oxford: Oxford University Press.

McMichael, P. (2009). A food regime genealogy. *Journal of Peasant Studies, 36*(1), 139–170.

McMichael, P. (2013). *Food regimes and agrarian questions*. Halifax: Fernwood Publishing.

Mol, A. P. J. (2001). *Globalization and environmental reform: The ecological modernization of the global economy*. Cambridge: MIT Press.

Molina, O., & Rhodes, M. (2007). The political economy of adjustment in mixed market economies: A study of Spain and Italy. In B. Hancké, M. Rhodes, & M. Thatcher (Eds.), *Beyond varieties of capitalism: Conflict, contradictions, and complementarities in the European economy* (pp. 223–252). Oxford: Oxford University Press.

OECD. (2001). *Multifunctionality: Towards an analytical framework*. Paris: OECD Publications Service.

OECD. (2006). *The new rural paradigm: Policies and governance*. Paris: OECD Publications Service.

OECD. (2016). *A new rural development paradigm for the 21st century: A toolkit for developing countries*. Paris: OECD Publishing. https://doi.org/10.1787/9789264252271-en.

Piore, M. (1979). *Birds of passage. Migrant labor and industrial societies*. Cambridge: Cambridge University Press.

Polanyi, K. (1944). *The great transformation*. New York: Farrar & Rinehart.

Portes, A., & Bach, R. L. (1985). *Latin journey. Cuban and Mexican immigrants in the United States*. Berkeley: University of California Press.

Portes, A., & Jensen, L. (1989). The enclave and the entrants: Patterns of ethnic enterprise in Miami before and Mariel. *American Sociological Review, 54*, 929–959.

Reich, M., Gordon, D. M., & Edwards, R. C. (1973). Dual labor markets: A theory of labor market segmentation. *American Economic Review, 63*(2), 359–365.

Reyneri, E. (2004). Immigrants in a segmented and often undeclared labour market. *Journal of Modern Italian Studies, 9*(1), 71–93.

Saitta, P. (2008). Tra struttura e funzione. Una critica degli approcci razionalisti in materia di immigrazione. *Studi Emigrazione, 169*, 135–158.

Sjaastad, L. (1962). The costs and returns of human migration. *Journal of Political Economy, 70*, 80–93.

Titmuss, R. M. (1963). *Essays on 'The welfare state'*. London: Unwin University Book.

Todaro, M. (1969). A model of labor migration and urban unemployment in less developed countries. *The American Economic Review, 59*, 138–148.

Van der Ploeg, J. D. (2008). *The new peasantries. Struggles for autonomy and sustainability in an era of empire and globalization*. London: Earthscan.

Van der Ploeg, J. D. (2013). *Peasants and the art of farming: A Chayanovian manifesto*. Winnipeg: Fernwood Publishing.

Van der Ploeg, J. D., & Marsden, T. (2008). *Unfolding webs: The dynamics of regional rural development*. Assen: Royal Van Gorcum.

Van der Ploeg, J. D., Henk Renting, H., Brunori, G., Knickel, K., Mannion, J., Marsden, T., De Roest, K., Sevilla-Guzmán, E., & Ventura, F. (2000). Rural development: From practices and policies towards theory. *Sociologia Ruralis, 40*(4), 391–408.

Van Huylenbroeck, G., & Durand, G. (Eds.). (2003). *Multifunctional agriculture: A new paradigm for European agriculture and rural development*. Aldershot/Burlington: Ashgate.

Waldinger, R. (1994). The making of an immigrant niche. *Intenational Migration Review, 28*(1), 3–30.

Waldinger, R., & Lichter, M. I. (2003). *How the other half works: Immigration and the social organization of labor*. Berkeley: University of California Press.

Wallerstein, I. (1979). *The capitalist world economy*. Cambridge: Cambridge University press.

Wilson, G. A. (2007). Multifunctional agriculture. In *A transition theory perspective*. Trowbridge: Cromwell Press.

Chapter 2
Restructuring of Agriculture and the Rural World in Mediterranean EU Countries

Agriculture and rural development represent critical domains for the economy, the society as well as the ecosystems of Euro-Mediterranean countries. Important changes and challenges have though reconfigured food production, natural resource management as well as rural livelihoods in recent decades in the region.

Main distinct but intertwined processes include: (i) Agricultural modernization and polarization; (ii) The restructuring of agri-food chains in the global market; (iii) The institutionalization of the agrarian world, including the role of the Common Agricultural Policy (CAP).

These processes have resulted in an increasing demand for lower-waged workers and the socio-economic marginalisation of rural communities, reducing the local attractiveness of agriculture and rural livelihoods. The reconfiguration of agricultural labour has resulted in a restructuring of its manpower, with a significant shift from family labour to a salaried, foreign one. It is within such framework that the consistent and growing presence of immigrants in rural areas and agricultural sector is to be assessed. The focus is on EUMed countries (Greece, Spain, and Italy), which present some specific and characterising features and dynamics.

2.1 A Focus on the Agrarian World of Mediterranean Europe

In recent decades the presence of migrants in rural areas has increased, stimulated by a growing demand for low-cost agricultural labour. In this chapter we will analyze the changes of the agrarian world and their links with migrations.

© The Author(s) 2020
M. Nori, D. Farinella, *Migration, Agriculture and Rural Development*, IMISCOE
Research Series, https://doi.org/10.1007/978-3-030-42863-1_2

Agriculture continues to play a strong role in rural areas of Mediterranean countries in the European Union (EUMed from here onwards[1]), as it defines social, environmental, economic as well as cultural identities. Agricultural products and rural tourism contribute consistently to national GDPs, and rural communities play a critical role in the management of biodiversity in Mediterranean ecosystems, where desertification is a threat.[2] A typical indicator exemplifying the relevance of agriculture in the EUMed compared to other European regions is that half of the agriculturally employed population and two-thirds of farm holdings in the EU-15 were concentrated in the European south (EU 2012).

Agriculture and rural societies in Europe have undergone critical changes and reconfiguration since WW II characterized by these three main distinct but intertwined processes:

1. Agricultural modernization and polarization;
2. The restructuring of agri-food chains in the global market;
3. The institutionalization of the agrarian world, including the role of the Common Agricultural Policy (CAP).

The substantial influx of immigrant communities in rural settings today is a consequence of this restructuring.

The modernization of agriculture has hinged on a market-oriented vision, which has promoted productivity over any other aspect of farming and rural development. Such processes have favoured investments mostly in areas with high potential for agriculture intensification, while those areas considered more marginal due to their agro-ecological features lagged in attracting political attention and financial investment. Polarization thus started to reshape the rural world in geographical as well as in socio-economic terms.

These dynamics accelerated in the nineties, when EUMed agriculture became deeply integrated into global agricultural food chains which encouraged specialised productions oriented to fresh consumption and processing. Tomatoes, oranges, strawberries, olives, fruits, wines, cheeses produced in southern Europe started to serve the increasingly growing global consumption demand. The resulting process was supported, amongst others, by public policies and funding through the Common Agriculture Policy (CAP). Today the top export markets for most of EUMed's fruits and vegetables are Germany, Austria, Switzerland, France, Sweden, and the UK, while wines and cheeses are largely exported to the US, Russia and China.

The incorporation of the EUMed agri-food into globalized chains has highly impacted farmers, who have found themselves in a subaltern role, facing high price competition, squeezed by decreasing prices dictated by the market, with reduced profit margins and eroded negotiation power in international trade systems. The

---

[1] Implying mainly Greece, Spain and Italy, though parts of Portugal and France pertain as well to the region.

[2] The Mediterranean region represents the second world biodiversity hotspot, and one of the regions most impacted by climate change according to UNEP (2010) and to IPCC (2014).

modern agricultural restyling has shifted socio-economic roles and relationships; many farmers ceasing their activities, while those remaining have been forced to cut down on production costs, including labour.

This sequence of events has resulted in an increasing demand for lower-waged workers, which has contributed to reducing the local attractiveness of agricultural work. The reconfiguration of agricultural labour has resulted in a restructuring of its manpower, with a significant shift from family-labour to externally-sourced, salaried work. In parallel there has been a shift from hiring a more local to a more foreign workforce. It is in this framework, where the presence of immigrants in farming activities is growing, that needs to be assessed.

This context applies specifically to EUMed countries where agricultural labour is normally temporary and precarious and requires workers to move according to seasonal agriculture demands, specifically for harvesting. The growing demand for a flexible, low-skilled and cheap labour and the decreasing interest shown by local populations explains why EUMed rural areas have become increasingly attractive for immigrants.

In this chapter, we will provide a general overview of these processes and the related changes and impacts in relation to the restructuring of agriculture and rural areas during recent decades. Our focus is on EUMed countries, which present some specific and characterising features when analysing these dynamics.

2.2   The Impacts of Agricultural Modernization

In traditional Mediterranean agricultural systems, the most typical labour configuration has been family-based work in the form of self-employment and informal labour from family members. EUMed countries have a long agrarian tradition whereby cultivated land, crops and livestock are based on the family farm and on its labour, though a system that aims at ensuring the production as well as the reproduction of the farming system. Specifically, for smallholders, as well as taking care of their farm production system, members of peasant households would also lend their work seasonally to larger farms. This model changed rapidly since the end of the World War, when European rural areas have undergone an important process of agricultural modernization, instilled also by the Community Agricultural Policy (CAP) (on this debate see Hervieu and Purseigle 2012; Arnalte-Alegre and Ortiz-Miranda 2013; Ortiz-Miranda et al. 2013; Agnoletti 2013; Angonelli and Emanueli 2016). This process implied the decline of the peasant agricultural model which hinged on multi-functionality, polyculture, self-consumption, and the redundancy of internal production factors, including family labour (Van der Ploeg 2013).

The modernization pull, towards a new market-orientation and market-integration of agricultural systems, has aimed to increase production and reduce costs (Van der Ploeg 2008, 2010). The aim to enhance agricultural "industrialization", through a "green revolution", has pushed for the intensification of production based on sectoral specialization, monoculture, standardisation and replicability; largely thanks to the application of tailored technologies as well as of dedicated chemical, agronomic and genetic sciences. The introduction of labour-saving machinery eventually triggered a crowding-out of rural populations. This in turn led to urbanization, to the growth of waged-labour and to the related transformation of the farmer from peasant to agricultural entrepreneur (Hervieu and Purseigle 2012). This also meant a "masculinization" of farming: women were increasingly marginalised from farming operations or assigned an auxiliary role in the farm economy, as they looked for employment in non-farm sectors (Saugeres 2002; Bharadwaj et al. 2013).

The restructuring of the agriculture world that has characterized recent economic development has contributed to the intensification of social and spatial differentiations in the rural world, with several relevant implications on farming and on farmers (Van der Ploeg 2008; Hervieu and Puseigle 2012). Agriculture has become increasingly integrated and dependent on market dynamics both upstream and downstream. Producers have lost their autonomy and have been forced to: acquire most production inputs (raw materials, technologies and other industrial inputs such as feed, seeds, pesticides, fertilizers, chemicals, genetics, oil and energy) on the market, and sell their products (the farm output) on the international market (Friedmann 2005).

The farming system increasingly lost its capacity to ensure the internal reproduction of its' means of production. Soil fertility is recovered through chemical fertilizers, plant and animal genetics elaborated elsewhere and acquired through the market, manpower either replaced by machines or scaled down to waged-labour. In this model, most of the farm's output is devoted to market exchanges. Generating income becomes necessary to purchase production inputs. Moreover, agricultural goods have become commodities in a global market, with high price volatility, hierarchical networks and decreasing returns to producers. As we will describe, all these factors have contributed to creating a subordinate position for producers, both in the value chains regulating distribution, and for the commercialization of their outputs.

These processes increase the phenomenon of the farm cost-price squeeze, caused by the growing gap between the Gross Value of Production (GVP) and the production costs incurred by the farmer. To maintain sufficient income, farmers have to increase the size of the enterprise in order to cut down on costs per unit. This eventually leads to a vicious circle whereby farm size is dictated by market prices and costs, and the farmers are squeezed by costs of production which increase faster than the price of their products (Moss 1992; Shield 2009).

The crisis of the agrarian world can be witnessed in the historical drop of farmer income, which eventually triggered the disqualification of agricultural work, the significant rates of land abandonment and the decline in number of farms, with related problems of rural exodus and "socio-economic desertification", as it will be assessed. Today rural areas in large parts of the EU are characterized by a declining and ageing population, low workforce availability and limited generational renewal. The low and decreasing percentage of young farmers in EU countries is considered a major problem for the future of agriculture. A short-term strategy relies on the increasing use of migrant workers to make up for the shortage of local labor.

However, in marginal territories, this process carries longer-term consequences for the reproduction of local societies. Furthermore, the abandonment of agriculture and land provokes a degradation of natural resources, the loss of ecological and cultural biodiversity, and a growth in regional disparities. Together these dynamics are seen to threaten the sustainability of agriculture, food systems and rural lifestyles (EU 2012; Zagata and Sutherland 2015; Nori 2017; SOFA 2018).

2.3   The Restructuring of Agri-Food Chains in the Global Market

The negative effects of agricultural modernization were amplified at the beginning of the 1990s, when neo-liberal processes of global restructuring of the agro-food supply chain contributed to unbalancing market relations, and increasing the power of large corporations through processes of unfair liberalization.

In economic theory, a supply chain includes different kinds of economic actors, operating in one or more phases of the chain and differing by size and economic power with varying degrees of relationships. The agri-food chain links producers to end consumers, and it consists of four distinct and consequent phases (Fig. 2.1):

1. Production of raw food, in which the companies operating in the primary sector are located (agriculture, livestock, fisheries). In this phase we find farmers, either independent or in cooperatives. Generally farmers are fragmented and small in size.
2. Transformation of raw materials, which involves attention to processing and manufacturing activities, be they industrial or artisanal. Food producers are of different sizes: small, medium (SME) and large enterprises, each one exhibiting and displaying different power and capacity to relate to other actors along the chain.
3. Packaging and labelling, which can be carried out both by the processing companies themselves or "purchased" by other service companies operating on the market.
4. Distribution and marketing, in which commercial and intermediary activities are located between the producer and the final consumer. Here we have a variegated universe from SMEs to large international groups and distributors, from small corner shops to supermarkets chains.

The two extremes of the chain show the largest degrees of risk and vulnerability. On the one hand the more intermediaries there are, the longer the supply chain and the lower the value and the control displayed by the producers. In addition, the longer the chain, the less control the consumers hold on it, since a high number of intermediary steps increases the information asymmetry to the detriment of the final user, despite the different measures of product traceability.

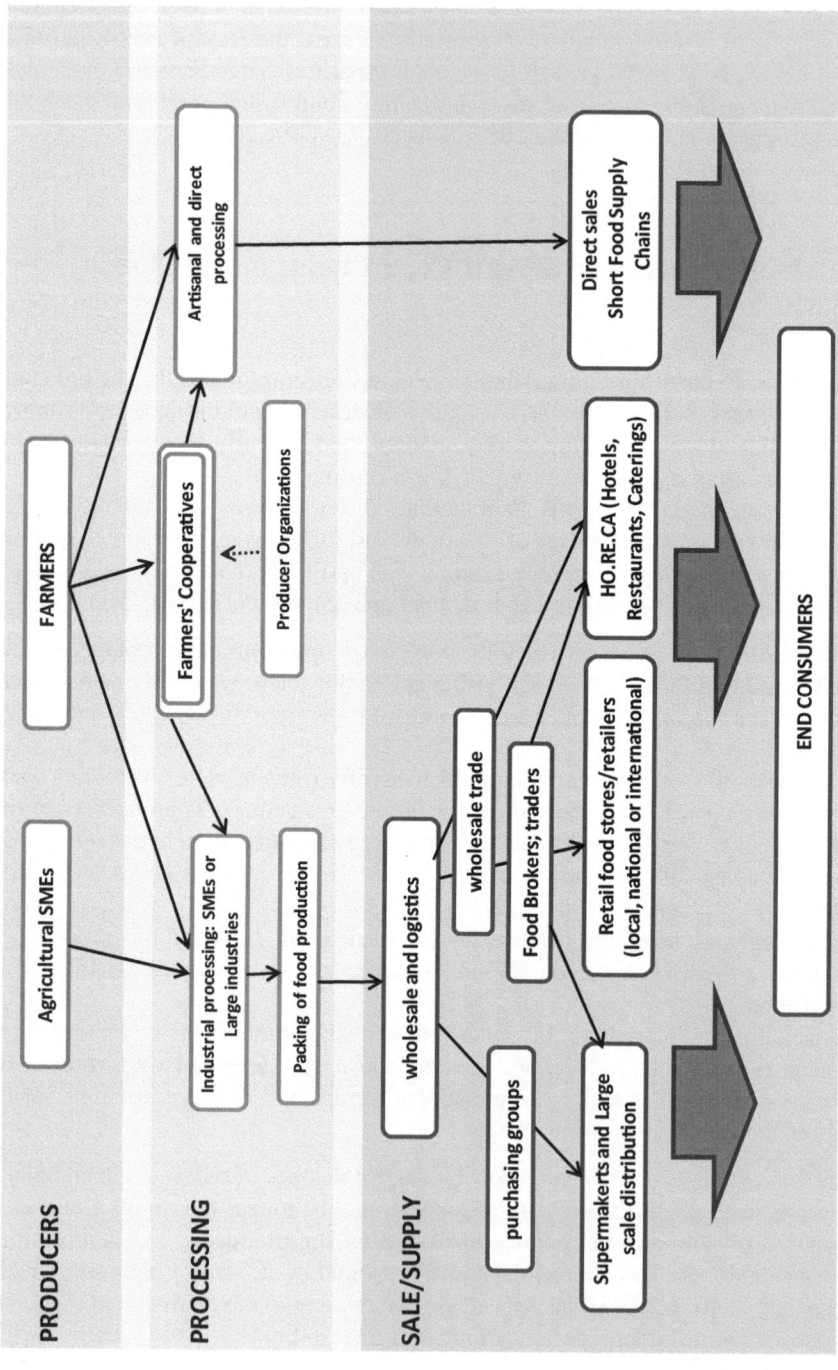

Fig. 2.1 The Agri-food chain. (Source: our elaboration)

On the other hand, a supply chain is short when there is a direct relationship between the producer (often also the processor) and the final consumer, without much distributive intermediation. Since the 2000s different types of Short Food Supply Chains (SFSC) and Alternative Food Networks (AFN) have spread as alternatives to mass production and large-scale organized distribution, in support of an agriculture embedded in the territory where the symbolic and relational values of food are also accounted for.

Box: Shorty Food Supply Chains and Alternative Food Network
The Short Food Supply Chain is an umbrella term to identify all alternative "short-circuits" that shifts from an 'industrial mode' of production and supply, engendering different relationships between producers and consumers (Mardsen et al. 2000):

> A key characteristic of short supply chains is their capacity to re-socialize or re-spatialize food, thereby allowing the consumer to make value-judgements about the relative desirability of foods based on of their own knowledge, experience, or perceived imagery. Commonly, such foods are defined either by the locality or even the specific farm where they are produced; and they serve to draw upon and enhance an image of the farm and/or region as a source of quality foods. 'Short' supply chains seek to redefine the producer-consumer relation by giving clear signals as to the origin of the food product. Short supply chains are also expressions of attempts (or struggles) by producers and consumers alike to match new types of supply and demand. Notable here are the additional identifiers which link price with quality criteria and the construction of quality. A common characteristic, however, is the emphasis upon the type of relationship between the producer and the consumer in these supply chains, and the role of this relationship in constructing value and meaning, rather than solely the type of product itself.

However, the Short Food Supply Chains represent a minority part of the global agri-food, which is characterized by an increasing concentration of retail corporate power over agricultural production. The restructuring of global value chains has been pushed by the neoliberal globalization (in 1990s), through several intertwined processes (Corrado et al. 2016a: 7):

> The incorporation of agricultural production in vertical food chains controlled by transnational corporations, the transformation from producer-driven to buyer-driven food chains (Burch and Lawrence 2007), the consolidation of retailer power through the supermarket revolution (Reardon et al. 2003; McMichael and Friedmann 2007), and the financialization of agricultural processes have all reshaped the global agri-food system and the connections between the global North and the global South.

As a result of this restructuring, distributors dispossess producers from leading the agri-food supply chains, whose hierarchy becomes vertically integrated, controlled and operated by transnational corporations that operate mainly in the distribution phase. These transnational corporations end up as intermediates to all relations that control the market, manipulating the functioning of agri-food chains to their advantage. Through an increasingly unfair distribution of risks, costs, and profits along the chain, food industries and retailers use their oligopolistic power of negotiation to impose price and contractual conditions on farmers, thus coming to weaken their managerial and economic capacities.

This "retailing revolution" is defined by agri-food chains progressively restructured in hierarchical networks, characterised by high price volatility, and decreasing returns to producers (McMichael and Friedmann 2007).

Using their asymmetric power in the distribution stage, large supermarket chains operate as *"food authorities* (Dixon 2007), imposing private standards upon agri-cultural production through retailer-driven agri-food supply chains" (Corrado et al. 2016a: 12), and control and affect as well other phases along the chain such as production, processing and consumption (Burch and Lawrence 2007). Van der Ploeg describes this phenomenon of lost autonomy from producers and consumers with the image of the "food empire" in which "it is becoming difficult, if not often impossi-ble, for farmers to sell food ingredients or for consumers to buy food outside of the circuits that they control" (Van der Ploeg 2010: 101).

Furthermore, the effective role and power of big transnational supermarket chains overcome the mere food chain, as these become able to influence the policies of nation states, pushing towards liberalizations that consolidate their power, both in developing countries and in those with an advanced economy. This has resulted in the crisis of small traders and retailers because the possibility to sell to the market passed to a few buying groups, that can impose their contractual conditions and prices (Vorley 2007).

Box: Expansion and Concentration of Agro-food Value Chains

Corrado et al. (2016a: 12) summarize the expansion and concentration of European agro-food value chain:

Europe's top 10 retail groups are headquartered in three countries: the UK, France and Germany. For example, in 2010, Carrefour (France) – Europe's largest retailer ahead of the Metro Group (Germany) and Tesco (UK) and second only to US-based Wal-Mart at the global level – employed 475,000 workers and had 15,600 company-operated or franchised stores in 34 countries across the world, with 57% of its turnover coming from outside France (Fritz 2011) [...] In Italy, large retailers' share of the food market grew from 44% in 1996 to 71% in 2011 (AGCM 2013). In Greece, the four largest retailers (three foreign chains and one national company) accounted for 55% of the sales and more than 80% of the profits of the national grocery retail market in 2009 (Skordili 2013). In Spain, big retailers controlled 63.7% of the food market in 2014 (ANGED 2014: 36). In Morocco, supermarket trade took off in the early 2000s, with the arrival of foreign direct investments, mainly by the French Auchan group.

Through the World Trade Organization's (WTO) negotiations of the General Agreement on Trade and Services (GATS) and other free-trade policies adopted amongst others also by the EU, the supermarket chains can buy agricultural products almost all over the world, regardless of the place of production, the seasons and the transportation costs. Including in countries where prices are lower due to less stringent environmental and labor regulation. The dual objective is to buy at the lowest price and to stimulate producers to keep their prices low (Vorley 2007; Gertel and Sippel 2014; Corrado et al. 2016b; Oxfam 2018). This process and the related squeeze affect affects agriculture all over, through an increasing exploitation of labour and land, with visible effects on the conditions of the environment, the quality of the food and the rights and conditions of workers.

Many researchers have investigated the different mechanisms through which distribution transnationals are able to buy at increasingly lower prices, transferring costs to farmers, who increasingly suffer from the agricultural squeeze. The supermarket chains impose on producers many "quality standards'" linked to different aspects of production (in particular, high-quantity, low prices, quality, packaging, environment and food safety) that marginalize small farmers and artisanal producers, for whom the adjustment to these parameters is often difficult and expensive (Burch et al. 2013; Burch and Lawrence 2013; Richards et al. 2013). This mechanism is even more aggravated by the system of "private labels", a way through which the supermarket chains buy agri-food products and distribute them under their own brand labels, turning into "food business operators" (Vorley 2007).

Farmers are forced into unfair contracts with unilateral conditions and retroactive unfair changes to working contracts or unjustified threat of termination of contracts,[3] and practices bordering illegality, especially when they are small and medium-sized enterprises (SMEs), with a subaltern and weak position in the supply chain. There have been many reports and campaigns denouncing the unbalanced functioning of a global agri-food supply chain, also in the case of EU countries.

> **Box: The Agricultural Squeeze in EU-Countries**
> In 2009 the European Commission confirmed the dramatic situation in which farmers saw their added value increasingly eroded to the advantage of the distribution phase (EU 2009: 7 ss.):
>
> Total value-added for the food supply chain in the EU25 in 2005 was ~€540 billion, i.e. 5.2% of the total value-added of the European economy. The agricultural sector represented 24% of this total, the food industry 33% and the distribution sector 43% (13% for wholesale and 30% for retail). The value-added of each sector is thus increasing moving downwards along the chain: In 2005, the food industry value-added was 1.4 bigger than the value-added of agriculture and the distribution sector was 1.3 bigger than the food industry. Agriculture value-added has declined over the 1995–2005 decade, with a 1.5% per year decrease. [...] in the meanwhile, the other sectors of the chain have grown over the period [...] consequently, the pattern of distribution of value-added across the food supply chain has significantly changed in the EU25 during the 1995–2005 decade. The share of agricultural industry has consistently decreased under the combined effect of its negative value-added growth

(continued)

---

[3]On these aspects see: European Commission, 29 January 2016, Report on unfair business-to-business trading practices in the food supply chain.

and the much more dynamic growth of the other sectors. The share of agriculture in food supply chain has decreased from 31% in 1995 (equal to the share of the food industry) to 24% in 2005 (with a food industry at 33%). The distribution sector has increased its share in the same period by 2% for the food wholesale sector (from 11% in 1995 to 13% in 2005 and by 3% for the food retail sector (from 27% to 30%).

The brunt of such impacts is typically borne by those holding a subordinate position within the value chain: on the one hand, farmers and their workers and on the other the final consumers. The general trend has been a loss of food sovereignty, decreased control on the quality of the production process and increasing dependence on unstable and volatile global market.

In a continuous race to the bottom, the farmers try to extract the lost value in the chain, making more use of two main productive factors, land and work. This leads to the exhaustion of the natural resopurce base on the one hand, and greater degrees of exploitation of the weaker and more precarious labor, often represented by immigrants, on the other.

2.4   The Institutionalization of the Agrarian World and the Role of the Common Agricultural Policy

The policy framework has therefore played a considerable and ambivalent role in the process of agricultural modernization and the oligopolistic restructuring of the global agri-food chain including in the EU with its prominent Common Agricultural Policy (CAP). CAP represents a main pillar of the European Union, and one of its main relevant policy axes; in 2018 it still engaged about 40% of the overall EU budget. The CAP was introduced in 1962, and for the first two decades it mainly spurred agricultural production within a framework of modernization of agriculture and the development of the global agri-food chain. This approach led to excess food supply and related market distortions, which eventually induced CAP reforms to better account for different aspects of European rural development within a more multifunctional perspective.

Overproduction, environmental problems, and consumer concerns for health and quality motivated CAP reforms through measures such as the reduction of price supports (through the 1992 MacSharry reform), cross-compliance with environmental objectives and support to agricultural multifunctionality and rural development (with Agenda 2000 programme), and the decoupling of direct payments from production according to certain conditions, whereby producers were no longer paid according to the quantities they produced, but based on the quality of the production process (2003 Fischler reform) (on the CAP reforms see, inter alia, Garzon 2006; Cuhna and Swinbank 2011; Swinnen 2015; Papadopoulos 2015b; Corrado et al. 2018).

In other words, payment is increasingly subject to compliance with the rules on environmental protection, food safety, animal and plant health and animal welfare, as well as with the obligation to maintain the land in good agricultural and ecological conditions. Payments are increasingly "rewards" which are greater for those farmers able to carry out measures of greening and agricultural biodiversity (for example with the preservation of native breeds, the diversification of crops, the maintenance of permanent grasslands and the care of the forest). A series of incentives are provided for those who work inland, in disadvantaged, remote and/or poorly connected areas, with a view to counter depopulation and abandonment, for example with incentives that encourage youth entrepreneurship, organic production and animal welfare practices.

> **Box: The New Rural Development Paradigm**
> According to Van der Ploeg et al. (2000: 392), a new rural development paradigm is taking place in both policy and practice, to contrast the negative effects of the modernization paradigm. Rural development (RD) is being "recognized as a multi-level process rooted in historical traditions". The focus is on re-embedding agriculture in the local society, in opposition to the tendency of modernization practices to segregate "agriculture" from the other rural activities. The RD paradigm is based on the idea that agriculture must be conceived as "multi-functional", producing not only agricultural commodities for the global market, but also services and collective goods. These are unique and non-transferable through markets, and include landscapes, natural values and agro-ecological biodiversity, local economy and social network to contrast rural abandonment. In this sense "many rural development experiences creating cohesion between activities, not only at farm level but also between different farms or farms and other rural activities, appear to be a crucial, strategic element. Particularly important are the (potential) synergies between local and regional eco-systems (Guzman Casado et al. 2000), specific farm styles, specific goods and services, localized food-chains and finally, specific social carriers and movements" (Van der Ploeg et al. 2000: 393). The territorial constraints become specific and non-imitable resources for place-based development paths. The local household farm is central to this process, with its ability to create value through economies that are alternative to neoliberal markets and embedded in relational and local circuits (what are defined as *nested markets*).

Following WTO agreements, CAP progressively moved towards stronger market orientation and agricultural sustainability, with an enhanced concern for quality processes and products (i.e., organic agriculture certifications and denominations of origin quality control). However, the related distorting effects mostly favoured food processors, the agrochemical industry, and large farms, but also export-oriented food traders and large retailers, with a controversial impact on developing countries. Some crops, territories, actors and companies have been more able than others to benefit from such schemes, with medium and large farms typically being favoured.

Similarly, EU support for producer organizations (POs) is criticised as having failed to enhance collective actions and reduce the fragmentation of farmers, while it favoured the cooperation of the most powerful and economically important stakeholders in the sector (Corrado et al. 2018).

While claiming support to small producers the role of the EU has often had the opposite effect by reorganizing agri-food chains in a neoliberal framework, which eventually undermined the power and the capacities of agricultural producers, whom have lost out in this process. Trade liberalization policies in recent years have reconfigured value chain dynamics through regulations concerning food safety, packaging, distribution and retailing, which eventually distorted power relations in favour of large industries and distribution corporations. Standards, certifications, and regulatory adjustments imposed by EU policies are often costly constraints and barriers to entry to markets that undermine the survival of small independent producers. As a result, farmers and rural producers have become the main shock absorbers of market risks resulting from the policy-assisted reconfiguration of agri-food value chains.

The modernisation of the agricultural sector resulted in an important polarisation of the territorial as well as social landscapes. On the one hand areas with higher potential for agriculture (ie. low plains, valley bottoms, coastal areas) have undergone intensification of agriculture production, while on the other hand more marginal settings where the potential for agricultural intensification is structurally limited, have witnessed a progressive abandonment.

Throughout the Mediterranean marginal communities, inhabiting mountainous, island territories or inner areas have carried the higher burden with entire territories depopulated, agricultural surfaces abandoned, and rural villages emptied through forms of socio-economic desertification. The climatic, financial and political crises that have characterised the last decade (and that are closely intertwined, as properly noted by Klein 2016) have compounded a polarised situation that was already quite stretched for Mediterranean agriculture and rural areas.

Overall the implications of such reconfiguration of agro-ecological and socio-economic landscapes have been dramatic. Family farming has become a decreasingly viable enterprise, while opportunities for agricultural workers have been jeopardised by the growing mechanisation on higher potential areas and by land abandonment in lower potential ones. Although each country has experienced different rates and modalities, such processes have altogether implied an important movement of populations out of rural areas.

Despite its relevant engagement, the EU's "rural welfare" scheme is increasingly criticized for its inability to offset the negative social and environmental trends affecting the EU agrarian world. Farmers in Europe increasingly rely on subsidy schemes, rural populations continue to decline, and remain socially and politically marginalised. Compared to neighboring urban areas, today the living conditions in EU rural areas are tougher, the quality of basic services and facilities are inferior and limited, and opportunities for employment and income are lower.

These features make living and working in the countryside an unattractive option for the local youth, who often tends to seek livelihood opportunities elsewhere.

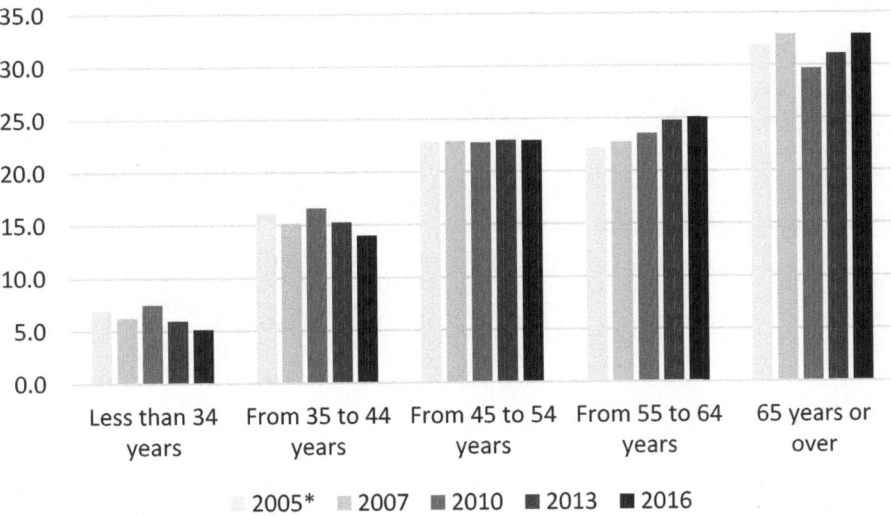

Fig. 2.2 Farmers by age (%) in EU-28. (Source: our re-elaboration on EUROSTAT, see https://ec.europa.eu/eurostat/web/agriculture/data/database. Map Legend: 2005 is on EU-27)

Similarly, to most rural areas throughout the world, a marking feature of the EU's countryside is the emigration of its rural youth, which leads in turn to the demographic aging of rural communities and problems of workforce availability and generational renewal in agricultural enterprises (refer to Fig. 2.2). In 2016 55% of EU-28 farmers are aged 55 years or more, and 32.8% are aged over 65. These values have increased compared to 2005, when the farmers aged 55 or more were the 54.1% and the farmers aged over 65 were 31.9%. The percentage of female farms remain still very low: 28.4% in 2016.

In the EUMed, agriculture is losing 2–3% of its active population per year. Today only one out of every ten farmers across the EUMed is younger than 35 years, while the percentage of the population aged over 65 represents more than 20% of those inhabiting rural areas (Table 2.1). Portugal leads the group with 22.7% of its rural population in this age group, followed by Greece (21.4%), Spain (21.1%), Italy (20.9%) and France (20.8%). Essentially, in the EUMed, agricultural labour force is older than in any other sector of the economy. These data lead to serious concerns about an increasingly ageing and dependent population in many rural areas, and structural consequences including land abandonment, depopulation, and lack of services which will further reduce the attractiveness of living in rural areas (Dollé 2011; Collantes and Pinilla 2011; Arnalte-Alegre and Ortiz-Miranda 2013; Collantes et al. 2014; Camarero and del Pino 2013; Campagne and Pecqueur 2014; Leavy and Hossain 2014; Papadopoulos 2015; Corrado et al. 2016b; Nori 2016; Farinella et al. 2017).

These dynamics have led to significant agrarian change over the last three decades, as the number of farms has steadily decreased, as it has, to a lesser extent,

Table 2.1 Basic statistics and trends of EUMed agriculture

Average farm size UAA (hectare)

	Greece	Spain	Italy	EU 28 average
1990	4.3	15.4	5,6	
2000	4.4	20.3	6.1	12.8[a]
2010	7.2	24	8	14.4 (EU 28)
2016	6.6	24.6	11	16.6 (EU 28)

Utilised agricultural area (UAA) (hectare)

	Greece	Spain	Italy	EU 28 average
1990	3.661.210	24.531.060	14.946.720	
2000	3.583.190	26.158.410	13.062.260	200.462.070[a]
2010	5.177.510	23.752.690	12.856.050	175.845.490
2016	4.553.830	23.229.750	12.598.160	173.338.550
% Decr/Incr. In 2000–16	27.1	−11.2	−3.6	−13.5 (y. 2003)

Farm holdings (quantity)

	Greece	Spain	Italy	EU 28 average
1990	850.140	1.593.640	2.664.550	
2000	817.060	1.287.420	2.153.720	15.669.410[a]
2010	723.060	989.800	1.620.880	12.245.700
2016	684.950	945.020	1.145.710	10.467.760
% 2000–16 rate drop	−16.2	−26.6	−46.8	−33.2

Employment in agriculture (total)

	Greece	Spain	Italy
2000	1.4 mil	2.4 mil	4 mil
2010	1.2 mil	2.2 mil	3.4 mil
% 1990–2010 Rate drop in agricultural employment	−15%	−8.7%	−14%

Employment in agriculture (% of total employment)

	Greece	Spain	Italy
2000	17.4	6.7	5.2
2010	12.4	4.2	3.8
2018	11.9	4.0	3.9

Labour force directly employed (LFE) - annual working unit

	Greece	Spain	Italy	EU 28 average
1990	680.330	1.143.350	1.923.990	
2000	587.480	1.077.730	1.364.920	14.229.940[a]
2010	429.520	888.970	953.790	9.943.950
2016	448.220	801.160	874.950	9.108.100
% 2000–16 Drop in LFE	−23.7	−25.7	−35.9	−36.0

Farm indicators by sex and age of the manager (%)

Variable	Year	Greece	Spain	Italy	EU 28
% Farmers aged over 65	2016	33.5	31.2	40.9	32.8
	2005	35.9	30.6	41.4	31.9
% Farmers aged between 55 and 64	2016	27.4	25.4	24.0	25.0
	2005	20.9	24.5	24.6	22.2

(continued)

Table 2.1 (continued)

Farm indicators by sex and age of the manager (%)

Variable	Year	Greece	Spain	Italy	EU 28
% Farmers aged until 34	2016	3.7	3.8	4.1	5.1
	2005	6.7	6.0	3.5	6.9
% Female farms	2016	27.5	22.5	31.5	28.4
	2005	25.2	19.0	27.9	26.3

Source: Our elaboration on Eurostat (https://ec.europa.eu/eurostat/web/agriculture/data/database) and ILOSTAT (2019). Map Legend: [a] = year 2003, because 2000 is not available for EU 28

the utilized agricultural area (UAA), while the average size of farms has grown (EU 2011, 2012, 2013, 2017).

While these trends show regional as well as global patterns, data and rates are though particularly high in the European context and specifically in its Mediterranean rims. In the EU-28, farm holdings declined by 40% between 2003 and 2016 and the farm size grew from 12.8 ha of UAA in 2003 to 16.6 ha in 2016. In the same period EU-28 countries lost much agricultural land (−13.5%), passing from approximately 200 million hectares in 2003 to 173 million in 2016. The directly employed agricultural labour force, calculated in annual working units, decreased by 36%.

Table 2.1 offers a general idea about these transformations for the EUMed countries, where the number of farms decreased consistently between 1990 and 2016, a reduction largely due to the drop in the number of small farms, while the size of average farms increased. These statistics are particularly worrying especially when thinking about the huge and longstanding political and financial investments of the CAP.

In **Greece** there were 684.950 agricultural holdings in 2016, a 16% drop with 132.110 farms ceasing their activity since 2000. Agricultural labour force also decreased by 15% from 1.4 million in 2000 to 1.2 million in 2010. During that same period labour force directly employed in agriculture dropped by 23.7%. The impact of the economic crisis that began in Greece in 2009, leading to the adoption of a stability program in collaboration with international lenders, had and is still carrying dramatic consequences on the Greek economy and society. Gross domestic product fell by about a fourth, while unemployment increased by almost 20% of the total workforce in few years (Eurostat 2015; ELSTAT 2015).

In such stretched setting agriculture remains an important source of livelihood for most rural areas. However the age of the farm heads is very high: in 2016, 33.5% of farmers are aged over 65 and 60.9% are aged over 55. Salaried labour in agriculture does not seem an appealing option for local workforce; most agricultural workers are thus foreign, and immigrants have played an important role in supporting the survival, expansion and modernisation of farms as well as in their resilience in the current crisis (Kasimis and Papadopolus 2013; Ragkos et al. 2015, 2016).

In **Spain** there were 945.020 agricultural holdings in 2016, a 26.6% drop compared to 2000. In the same decade the land utilised for agricultural purposes decreased by 11.2%, while labour force directly employed in agriculture dropped by 25.7%. During that same time span the average farm size increased, passing from 15.4 (hectares/farm) in 1990 to 24.6 in 2016, while the average age of farmers slightly increased: in the 2005, 55.1% of farmers was aged over 55, in 2016 this is the 56.6%: just under one farmer every three is over 65 years old.

Between 2000 and 2010 the employment in agriculture passed from 2.4 million to 2.2 million (about 10% of the economically active population). Since the 1980s and 1990s, the development of intensive agriculture started relying on migrant workers, predominantly employed in an irregular manner through informal networks. Since the 2000s, agricultural work relations were formalized through mechanisms that aimed to serve the needs of a very flexible industry that had become increasingly specialized (Ortiz-Miranda et al. 2013; Corrado et al. 2016b).

In **Italy** there were 1.145.710 agricultural holdings in 2016, the third largest amount within the EU-28, after Romania and Poland. In 2000 the number of holdings was 2.153 million: in 16 years about 46.8% of the farms had ceased their activity. During the same period, the average size farm almost doubled, passing from 6.1 hectare/farm in 2000 to 11 hectare/farm in 2016. Today Italy has the third highest percentage of farmers aged over 65: 40.9% in 2016, after Romania (44.3%) and Cyprus (44.6%).

This process of concentration and modernization is also evident when looking at the decrease in agricultural work: the amount of persons working in agriculture dropped of about 14% in 2000–2010, from 4 to 3,four million; the labour force directly employed, calculated in annual working unit, dropped of 35.9% in 2000–2016. If we consider that in 2010 the agricultural labour force still represented 14% of the economically active population, it can be inferred that an important proportion of work is still informal, and it is carried out, as we will see, by migrant workers. A Caritas report indicated in 2014:

> Italian agriculture products are in the hands of foreign workers, accounting for about 25% of the total number of employment days in the food industry. (. . .) Foreign workers are contributing in a structural and critical way to the country's agricultural economy and are a much-needed component in ensuring the excellence of Italian food in the world.

Data from INEA seems to confirm these indications (refer to Fig. 2.3).

A comprehensively critical assessment of the CAP and of European Policies today would recognise their contribution in consolidating, and to an extent even widening, sectoral, social, generational and territorial inequalities. CAP has provided proportional advantages to larger farms and companies, higher-potential areas, intensive production systems and specialized agricultural enclaves. Conversely, and as a consequence, family farming and extensive agricultural systems have undergone dramatic processes of abandonment. The arrival of immigrants in rural areas has enabled tackling these dynamics, by countering the demographic decline and matching the demand for low-cost and flexible labour (Kasimis et al. 2010; Colloca and Corrado 2013; Caruso and Corrado 2015; Nori 2018; Farinella et al. 2017; ENRD 2018).

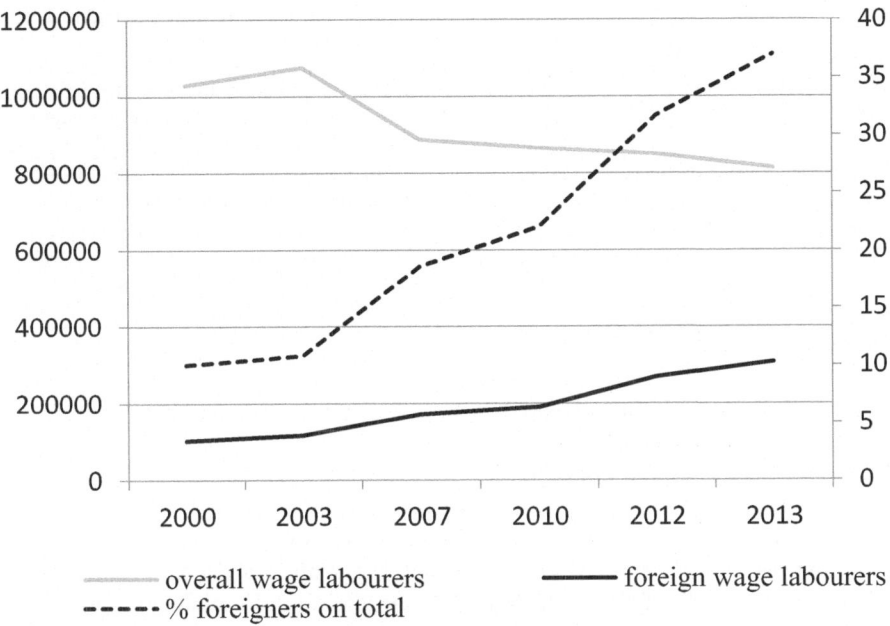

Fig. 2.3 Foreign workers in Italian agriculture (years 2000–2013). (Source: own elaboration on INEA data 2014)

In order to redress these dynamics recent CAP reforms allocate more emphasis on the wider rural context in which farming operates and its role in managing environmental, climatic as well as social matters. These principles inspired the CAP 2013 reform and are also present in the documents introducing the forthcoming one in 2020. In the EC communication *The Future of Food and Farming* (2017) specific mention is made of "generational renewal that should become a priority in a new policy framework" and tailored schemes that must be developed to "reflect the specific needs of young farmers" (EU 2017: 23). Moreover in the document there is an emphasis on that "the future CAP could play a larger role in addressing the root causes of migration" and that "the CAP can play a role in helping to settle and integrate legal migrants, refugees in particular, into rural communities".

2.5   Conclusions

In this chapter we highlighted the changes and dynamics that have affected European agriculture in recent decades, with a focus on the related implications and impacts for small-scale farmers and for marginal rural settings (Table 2.2). The processes of agricultural modernization and the oligopolistic restructuring of global agri-food

Table 2.2 The relationships between the policy framework and the shaping of the field reality

Process	Impacts	Implications
Modernization	Territorial polarization	Intensive agriculture in high-potential areas; abandonment lower potential ones; youth exodus
Value-chain restructuring	Oligopolistic control	Loss of food sovereignty and quality of the production system; agricultural squeeze
Rural development paradigm	Contrasting socio-economic and ecological downgrade	Sinergy between agriculture and territory; multi-functionality; local collective goods and agro-ecological services; local and alternative supply chain

Source: our elaboration

chains have generated socio-spatial marginalization, in the national territories as much as in the markets.

The consequences are borne by the whole society – producers, consumers, and citizens alike – whose capacity to influence food sovereignty, production quality and worker's rights has decreased. Despite important policy investments, several rural areas experience problems of environmental degradation and abandonment, thus adding to problems of desertification that are affecting different portions of the EUMed region.

Since the 2000s, a new rural development paradigm is emerging to contrast the modernization one, with a view to buffer the socio-economic downgrade of rural regions. This focuses on the idea to re-embed agriculture practices in the local society and ecosystems, enhancing the natural multi-functionality of agriculture and its capacity to produce also services and collective goods such as biodiversity, landscapes, tacit knowledge, local economy and social relations. These principles which would eventually look into the quality of the production process and also include the conditions of its workers has though not yet materialized in policy terms.

Through its economic, social and ecological implications the consequences of agricultural restructuring represent critical political issues. On the one hand the most fragile, inland and mountain territories face growing degrees of marginalization, increasingly emptied by demographic decline and land abandonment, facing socio-economic desertification. On the other hand, territories with higher agricultural potentials suffer the burden of encroaching urbanization, agricultural intensification, pollution and overexploitation, with often irreversible outcomes. Risks and hazards associated to unsustainable management of natural resources have characterized recent important, tragic events all over the EUMed region. Farmers, the most critical but the weakest component of the supply chain, bear the brunt of the costs of reducing earnings, which pushes towards reducing production costs and often translates into exploitative regimes, including workers. These workers are often, and increasingly, of foreign origin. Their presence significantly contributes to redress the social and economic mismatch affecting EUMed agrarian worlds, by filling the gaps left by local populations. The living and working conditions of rural immigrants are strictly regulated by a policy framework which provides them with limited and ineffective rights. The principles inspiring the CAP and its generous hand-outs are informed by animal welfare, organic production, food safety much more then by their care for the rights and conditions of agricultural workers (Corrado et al. 2018).

In the following chapters we try addressing some of these critical issues in relation to the migrant workers in rural regions.

References

AGCM (Autorità Garante della Concorrenza e del Mercato) (2013). Indagine conoscitiva sul settore della GDO. Rome: AGCM

Agnoletti, M. (Ed.). (2013). *Italian historical rural landscapes. Cultural values for the environment and rural development*. Dordrecht: Springer.

Agnoletti, M., & Emanueli, F. (Eds.). (2016). *Biocultural diversity in Europe*. Berlin: Springer.

ANGED (Asociación Nacional Grandes de Empresas de Distribución) (2014). Informe annual, 2014. Madrid: ANGED.

Arnalte-Alegre, E., & Ortiz-Miranda, D. (2013). The 'Southern Model' of European agriculture revisited: Continuities and dynamics. In D. Ortiz-Miranda, A. Moragues-Faus, & E. Arnalte-Alegre (Eds.), *Agriculture in Mediterranean Europe: Between old and new paradigms* (Research in rural sociology and development, Vol. 19, pp. 37–74). Bradford: Emerald Group Publishing Limited.

Bharadwaj, L., Findeis, J. L., & Chintawar, S. (2013). Motivations to work off-farm among US farm women. *Journal of Socio-Economics, 45*, 71–77. https://doi.org/10.1016/j.socec.2013.04.002.

Burch, D., & Lawrence, G. (Eds.). (2007). *Supermarkets and Agri-food supply chains. Transformations in the production and consumption of foods*. Cheltenham: Edward Elgar.

Burch, D., & Lawrence, G. (2013). Financialization in agri-food supply chains: Private equity and the transformation of the retail sector. *Agriculture and Human Values, 30*, 247–258. https://doi.org/10.1007/s10460-012-9413-7.

Burch, D., Dixon, J., & Lawrence, G. (2013). Introduction to symposium on the changing role of supermarkets in global supply chains. From seedling to supermarket: Agri-food supply chains in transition. *Agriculture and Human Values, 30*, 215–224. https://doi.org/10.1007/s10460-012-9410-x.

Camarero, L., & Del Pino, J. A. (2013). Impacts of ageing on socioeconomic sustainability in Spanish rural areas. In M. Gather (Ed.), *Proceedings of the 2nd Eurufu scientific conference: Education, local economy and job opportunities in rural areas* (pp. 79–89). Erfurt: Institut Verkehrund Raum.

Campagne, P., & Pecqueur, B. (2014). *Le développement territorial: une réponse émergente à la mondialisation*. Paris: Fondation Charles Léopold Mayer.

Caritas. (2014). *Dossier statistico immigrazione 2014* (Rapporto Unar). Roma: Portale Integrazione Migranti.

Caruso, F., & Corrado, A. (2015). Migrazioni e lavoro agricolo: un confronto tra Italia e Spagna in tempi di crisi. In M. Colucci & S. Gallo (Eds.), *Tempo di cambiare: Rapporto 2015 sulle migrazioni interne in Italia*. Roma: Donzelli.

Centre for European Policy Studies. (2015). In J. Swinnen (Ed.), *The political economy of the 2014–2020 common agricultural policy. An imperfect storm*. London: Rowman& Littlefield International.

Collantes, F., & Pinilla, V. (2011). *Peaceful surrender: The depopulation of rural Spain in the twentieth century*. Newcastle-upon-Tyne: Cambridge Scholars Publishing.

Collantes, F., Pinilla, V., Sàez, L. A., & Silvestre, J. (2014). Reducing depopulation in rural Spain. *Population, Space and Place, 20*, 606–621. https://doi.org/10.1002/psp.1797.

Colloca, C., & Corrado, A. (Eds.). (2013). *La globalizzazione delle campagne* (Migranti e società rurali nel Sud Italia). Milano: FrancoAngeli.

Corrado, A., De Castro, C., & Perrotta, D. (2016a). Introduction. Cheap food, cheap labour, high profits. Agriculture and mobility in the Mediterranean migration and agriculture. In A. Corrado, C. De Castro, & D. Perrotta (Eds.), *Mobility and change in the Mediterranean area* (pp. 1–46). London/New York: Routledge.

Corrado, A., De Castro, C., & Perrotta, D. (Eds.). (2016b). *Mobility and change in the Mediterranean area*. London/New York: Routledge.

Corrado, A., Palumbo L., Caruso F. S., M.Lo Cascio., Nori M., Traindafyllidou, A., (2018). Is Italian agriculture a 'Pull Factor' for Irregular Migration – and, if so, why?, Open Society Foundations. https://www.opensocietyfoundations.org/sites/default/files/is-italian-agriculture-a-pull-factor-for-irregular-migration-20181205.pdf. Accessed 20 Jan 2019.

Cuhna, A., & Swinbank, A. (2011). *An inside view of the CAP reform process, explaining the MacSharry, agenda 2000 and Fischler reforms*. Oxford: Oxford University Press.

Dixon, J. (2007). Supermarkets as new food authorities. In D. Burch & G. Lawrence (Eds.), *Supermarkets and agri-food supply chains: Transformations in the production and consumption of foods* (pp. 29–50). Cheltenham: Edward Elgar.

Dollé, V. (2011). *Food security and agriculture in the Mediterranean: Crisis scenario and prospects for 2030*. In *Tomorrow the Mediterranean* (*Scenarios and projections for 2030*). Paris: IPEMED.

EC. (2009). *A better functioning food supply chain in Europe. A view to promoting quality and safe food products at affordable prices*. Bruxelles: European Commission. http://europa.eu/rapid/press-release_MEMO-09-483_en.htm?locale=en.

ELSTAT. (2015). *Greek statistical authority*. http://tinyurl.com/jejuhgs

ENRD. (2018). *Strategy for inner areas*. Italy: European Network for Rural Development. https://enrd.ec.europa.eu/sites/enrd/files/tg_smart-villages_case-study_it_0.pdf.

EU. (2011). *Impact assessment of the CAP reform*. Bruxelles: European Commission. https://ec.europa.eu/agriculture/sites/agriculture/files/policy-perspectives/impact-assessment/cap-towards-2020/report/full-text_en.pdf.

EU. (2012). *Rural development in the EU statistical and economic information*. Bruxelles: EC DG Agriculture. https://ec.europa.eu/agriculture/statistics/rural-development_en.

EU. (2013). *The CAP towards 2020: Meeting the food, natural resources and territorial challenges of the future*. Bruxelles: EC DG Agriculture. http://ec.europa.eu/dorie/fileDownload.do;jsessionid=nxRcuiAAqmbkiw5FbJPj4u6gT9cczLOYbz9mDWZDxNrd1Pam0cYs!1583997504?docId=1336699&cardId=1336698.

EU. (2017). *The future of food and farming*. Communication from the Commission to the European Parliament, the Council, the European Economic and Social Committee and The Committee of the Regions. Draft document introducing the CAP 2020 Reform https://ec.europa.eu/agriculture/sites/agriculture/files/future-of-cap/future_of_food_and_farming_communication_en.pdf

Eurostat. (2015). *Agriculture, forestry and fishery statistics, 2015th edition*. Luxembourg: Eurostat.

Farinella, D., Nori, M., & Ragkos, A. (2017). Change in Euro-Mediterranean pastoralism: Which opportunities for rural development and generational renewal? In C. Porqueddu, A. Franca, G. Molle, G. Peratoner, & A. Hokings (Eds.), *Grassland reources for extensive farming systems in marginal lands: major drivers and future scenarios* (Vol. 22, pp. 23–36). Wageningen: Grassland Science in Europe. Wageningen Academic Publishers.

Friedmann, H. (2005). From colonialism to green capitalism: Social movements and the emergence of food regimes. In H. Buttel, P. McMichael, & P. Oxford (Eds.), *New directions in the sociology of global development* (pp. 229–267). Oxford: Elsevier.

Fritz, T. (2011). *Globalising hunger: Food security and the EU's common agricultural policy*. Berlin: Verlag.

Garzon, I. (2006). *Reforming the CAP: History of a paradigm change*. Houndmills/Basingstoke/Hampshire: Palgrave Macmillan.

Gertel, J., & Sippel, S. R. (Eds.). (2014). *Seasonal workers in Mediterranean agriculture. The social costs of eating fresh*. London: Routledge.

Guzman Casado, G. I., González de Molina, M., & Sevilla Guzmán, E. (2000). *Introduccion a la Agroecologia Como Desarrollo Rural Sostenible*. Madrid: Ediciones Mundi-Prensa.

Hervieu, B., & Purseigle, F. (2012). *Sociologie des mondes agricoles*. Paris: Armand Colin.

ILOSTAT. (2019). *Employment in agriculture*. https://data.worldbank.org/indicator/SL.AGR. EMPL.ZS?locations=IT-ES-GR&view=chart. Accessed 20 Jan 2019.

INEA. (2014). *Annuario dell'agricoltura Italiana, 2013*. Roma: Istituto Nazionale di Economia Agraria.

IPCC. (2014). *Fifth Assessment Report, Intergovernmental Panel on Climate Change*. www.ipcc. ch/report/ar5/

Kasimis, C., & Papadopoulos, A. G. (2013). Rural transformations and family farming in contemporary Greece. In D. Ortiz-Miranda, A. Moragues-Faus, & E. Arnalte-Alegre (Eds.), *Agriculture in Mediterranean Europe: Between old and new paradigms* (Research in rural sociology and development, Vol. 19, pp. 283–293). Bradford: Emerald Group Publishing Limited.

Kasimis, C., Papadopoulos, A. G., & Pappas, C. (2010). Gaining from rural migrants: Migrant employment strategies and socioeconomic implications for rural labour markets. *Sociologia Ruralis, 50*(3), 258–276. https://doi.org/10.1111/j.1467-9523.2010.00515.x.

Klein, N. (2016). Let them drown: The violence of othering in a warming world. *London Review of Books, 38*(11), 11–14. 5421.

Leavy, J., & Hossain, N. (2014). Who wants to farm? Youth aspirations, opportunities and rising food prices. IDS Working Paper, 439, 1–44, https://doi.org/10.1111/j.2040-0209.2014.00439.x

Marsden, T., Banks, J., & Bristow, G. (2000). Food supply chain approaches: Exploring their role in rural development. *Sociologia Ruralis, 40*, 424–438. https://doi.org/10.1111/1467-9523.00158.

McMichael, P., & Friedmann, H. (2007). Situating the 'retailing revolution. In D. Burch & G. Lawrence (Eds.), *Supermarkets and Agri-food supply chains: Transformations in the production and consumption of foods* (pp. 291–319). Cheltenham: Edward Elgar.

Moss, C. B. (1992). The cost-price squeeze in agriculture: An application of Cointegration. *Review of Agricultural Economics, 14*, 205–213.

Nori, M. (2016). Shifting transhumances: Migrations patterns in Mediterranean pastoralism. In CIHEAM, Watch letter 36. *Crise et resilience en la Mediterranee*. Montpellier. www.iamb.it/ share/integra_"les_lib/"les/WL36.pdf

Nori, M. (2017). *Immigrant shepherds in Southern Europe*. E-paper, Heinrich Böll Stiftung Foundation. https://www.boell.de/en/agriculture-food-production-and-labour-migration-south ern-europe

Nori, M. (2018). Agriculture and rural territories in the Mediterranean: The case for mountainous communities. In Mediterra (Ed.), *Inclusion and migration challenges around the Mediterranean*. Paris: CIHEAM.

Ortiz-Miranda, D., Moragues-Faus, A., & Arnalte-Alegre, E. (2013). *Agriculture in Mediterranean Europe: Between old and new paradigms* (Research in Rural Sociology and Development, Vol. 19). Bingley: Emerald Group Publishing.

Oxfam. (2018). *Human suffering in Italy's agricultural value chain*. Oxfam International & Terra.

Papadopoulos, A. G., (2015). *In what way is Greek family farming defying the economic crisis?* Ağrı, 43. https://agriregionieuropa.univpm.it/it/content/article/31/43/what-way-greek-family-farming-defying-economic-crisis.

Papadopoulos, A. G. (2015b). The impact of the CAP on agriculture and rural areas of EU member states. *Agrarian South: Journal of Political Economy, 4*(1), 22–53. https://doi.org/10.1177/2277976015574054.

Ragkos, A., Theodoridis, A., Fachouridis, A., & Batzios, C. (2015). Dairy farmers' strategies against the crisis and the economic performance of farms. *Procedia Economics and Finance, 33*, 518–527.

Ragkos, A., Koutsou, S., & Manousidis, T. (2016). In search of strategies to face the economic crisis: Evidence from Greek farms. *South European Society and Politics, 21*, 319–337. doi.org/10.1080/13608746.2016.1164916.

Reardon, T., Timmer, C. P., Barrett, C. B., & Berdegue, J. (2003). The rise of supermarkets in Africa, Asia and Latin America. *American Journal of Agricultural Economics, 85*(5), 1140–1146.

Richards, C., Bjørkhaug, H., Lawrence, G., & Hickman, G. (2013). Retailer-driven agricultural restructuring—Australia, the UK and Norway in comparison. *Agriculture and Human Values, 30*, 235–245. https://doi.org/10.1007/s10460-019-09917-2.

Saugeres, L. (2002). Of tractors and men: Masculinity, technology and power in a French farming community. *Sociologia Ruralis, 42*, 143–159. https://doi.org/10.1111/1467-9523.00207.

Shield Dennis, A. (2009). *The farm Price-Cost squeeze and U.S. farm policy.* Congressional Research Service. 7-5700. www.crs.gov.

Skordili, S. (2013). "The sojourn of Aldi in Greece". Journal of Business and Retail Management Research, 8, 1, pp. 68–80.

SOFA. (2018). State of Food and Agriculture 2018 on 'Migration, agriculture and rural development'. http://www.fao.org/3/I9549EN/i9549en.pdf

UNEP. (2010). *Security in the Horn of Africa: The implications of a drier, hotter and more crowded future.* Nairobi: Global Environmental Alert Service, UN Environmental Programme.

Van der Ploeg, J. D. (2008). *The new peasantries: Struggles for autonomy and sustainability in an era of empire and globalization.* London: Earthscan.

Van der Ploeg, J. D. (2010). The food crisis, industrialized farming and the imperial regime. *Journal of Agrarian Change, 10*(1), 98–106. https://doi.org/10.1111/j.1471-0366.2009.00251.x.

Van der Ploeg, J. D. (2013). *Peasants and the art of farming: A chayanovian manifesto.* Winnipeg: Fernwood Publishing.

Van der Ploeg, J. D., Renting, H., Brunori, G., Knickel, K., Mannion, J., Mardsen, T., De Roest, K., Sevilla-Guzmán, E., & Ventura, F. (2000). rural development: from practices and policies towards theory. *Sociologia Ruralis, 40*(4), 391–408.

Vorley, B. (2007). Supermarkets and agri-food supply chains in Europe: Partnership and protest. In D. Burch & G. Lawrence (Eds.), *Supermarkets and Agri-food supply chains: Transformations in the production and consumption of foods* (pp. 245–269). Cheltenham: Edward Elgar.

Zagata, L., & Sutherland, L. A. (2015). Deconstructing the 'young farmer problem in Europe': Towards a research agenda. *Journal of Rural Studies, 38*, 39–51.

Chapter 3
Mobility and Migrations in the Rural Areas of Mediterranean EU Countries

This chapter focuses on the ambivalent nature of contemporary migrations in European rural areas. The growing presence of immigrants in these areas is a direct result of the restructuring of agriculture and global agri-food chains. Evidence indicates that while agricultural work and rural settings are decreasingly attractive to local populations, they represent a favourable environment to international newcomers, due to the higher chances to access livelihood resources. The non-visibility and informality that characterise rural settings and agricultural work arrangements provide on the one side opportunities for employment, while also fostering illegal labour practices and situations of harsh exploitation.

The specificities of the Mediterranean migration model are assessed accordingly, followed by a more in-depth analysis of the agricultural sector and the broader rural world in Greece, Spain and Italy.

3.1 Introduction

This chapter focuses on the ambivalent nature of contemporary agricultural migration in European rural areas. Rural immigrants can be viewed as either seasonal agricultural workers or as new citizens offsetting declining local populations and revitalizing the rural world.

The framework described in Chap. 2 detailed the reconfiguration of EU agriculture and the important implications on its agricultural workforce, which in recent decades has shifted from family labour to a salaried, foreign one. In the context of a progressive rural exodus of local populations, the relative proportion of immigrants has been raising rapidly in European farming and rural realities.

Evidence indicates that while agricultural work and rural settings are decreasingly attractive to local populations, they represent a favourable environment to international newcomers, due to easier chances to access livelihood resources compared to urban areas. Furthermore rural realities offer degrees of non-visibility and

© The Author(s) 2020

M. Nori, D. Farinella, *Migration, Agriculture and Rural Development*, IMISCOE Research Series, https://doi.org/10.1007/978-3-030-42863-1_3

informality, which help accommodate the needs of several immigrants. On the other hand, these same realities foster illegal labour practices and situations of harsh exploitation, as will be analysed subsequently.

Apart from filling the gaps in agricultural labour markets immigrants play multifunctional roles in rural settings, and their contributions are often vital to agricultural farms and rural societies, which in the last decades have undergone intense economic, demographic as well as socio-cultural decline.

While the presence of immigrants is visible throughout rural areas in Europe, it is particularly visible on the Mediterranean rim, where the demand for agricultural labour remains high while the decline and ageing of rural populations are particularly worrying.

Typically associated with emigration, the EUMed regions have turned within a few decades into an area of transition. They have eventually become a destination in its own right. Immigration to rural areas of the EUMed started in the 1980s and has expanded ever since. During the recent financial crisis, intense immigration has represented a key factor of resilience for the agricultural sector and the rural world, as it has enabled many farms, rural villages and agricultural enterprises to remain alive and productive throughout difficult times.

This chapter starts by providing a basic understanding of the growing centrality of immigrant workers in agriculture; the specificities of the Mediterranean migration model are then assessed, followed by a more in-depth analysis of the agricultural sector and the broader rural world in Greece, Spain and Italy. In next chapters we will further focus on the more marginal, isolated, remote areas of the EUMed where the contributions of immigrants are critical for the sustainability and reproduction of local rural societies.

3.2   Shifting Human Landscapes: The Growth of Immigration in Agriculture

According to Martin (2016: IX):

> Agriculture, the production of food and fiber on farms, employs a third of the world's workers, more than any other industry. An increasing share of the workers employed in industrial-country agriculture are hired or wage workers, and many of these are international migrants from poorer countries.

This phenomenon does not only apply only to industrialized Western countries, but it seems to be widespread throughout the globe as foreign workers are frequently, and increasingly, part of the agricultural workforce even in poorer and middle-income countries (OECD 2018; Zuccotti et al. 2018; SOFA 2018; UN DESA 2017; ILO 2015).

Box: Agricultural Migrant Labour in the US and Canada (SOFA 2018)
Foreign labour constitutes the backbone of agricultural production in Canada and the United States. The agricultural sector in Canada relies heavily on the labour of temporary migrant agricultural workers. Roughly 75% of labour gaps in the agricultural sector is filled by this group. Migrant workers have played a fundamental role in helping the country's horticultural industry to compete in the global food economy. In fact, evidence shows a direct link between the growth of Canadian horticultural exports and the rising number of migrant workers.

In the United States approximately 70% of farm workers are Mexican (the data is almost 90% in agriculture intensive California). When the inflow of Mexican immigrant workers decreased during the 2002–2014 period due to stricter border controls, shortages in agricultural labour resulted in significant losses to American farmers. Calculated revenue losses amounted to USD 3 billion for each year between 2002 and 2014.

Europe is not an exception to this and today, in many parts of the EU, immigrant workers (with or without legal status) constitute a consistent portion of the agricultural labour force. Disentangling the critical relationships between the conditions of agricultural work, rural development paradigms, labour markets and migration policies, represents thus a necessary step to understand ongoing dynamics.

On the one hand, farmers are increasingly squeezed by price competition and thus pushed towards consistent cuts in production costs, which eventually leads them to hire a cheap and flexible labour through an immigrant workforce. On the other hand, agriculture provides immigrants with better access to affordable basic resources like food and shelter and accessible employment and income-earning opportunities for low-skilled people originating from poorer regions. Rural areas and the agricultural sector also offer degrees of non-visibility and informality that help accommodate the needs of irregular migrant citizens and workers. These informal features might represent critical constraints to the integration of foreign workers in the agrarian world, contributing to undermining the social acceptance as well as the sustainability of agricultural systems (Gertel and Sippel 2014; Corrado et al. 2016; Farinella et al. 2017). These features are particularly specific to the EUMed context where, together with domestic work, agriculture is the main sector recruiting migrants without regular contracts (OECD 2012).

National governments respond to farm labour shortages through forms that range from; high tolerance of informal work in agriculture (meaning there may be little surveillance of immigration status in areas and sectors where there is a high labour demand) to increased attention to legal status once the harvest season is over. Policies that regulate the entry of seasonal workers ("quota policies"), and their ambiguous enforcement, have generated a lot of controversy and conflict, as these are mostly oriented to favour and protect local farms rather than their workers, of immigrant origin. Policies are not designed to counter the agricultural squeeze of farmers as a consequence of the global mechanisms of agri-food supply chains; rather, they incentivize agricultural companies to exploit workers, increasing control, surveillance and blackmail mechanisms. According to Martin (2016: X).

> Many governments enact protective labour laws after particular incidents involving farm worker protests or injuries. Few have policies to encourage farm employers to abide by these laws and raise labor standards and productivity over time so that agriculture provides higher-wage and safer jobs for more skilled workers. Instead, many governments tolerate unauthorized workers and do not adequately protect guest workers, leading to substandard housing and other deficiencies.

Informality is a tolerated factor and is produced and reproduced through labour policies that help farms contain production costs, compressing "living labour" and consequently wages. Apart from the high level of informality and precariousness that characterize several agrarian settings, immigrant workers are also difficult to reach for activists, civil society, institutional controls and researchers alike for several social, cultural as well as logistical matters, including the effective awareness of their rights.

3.3   A Mediterranean Model of Migration

In geo-global politics, southern Europe plays the role of a semi-peripheral zone (Arrighi 1985). From the beginning of the twentieth century until the 1970s, southern European countries were places of emigration, characterized by internal flows, from rural areas to urbanized and industrial ones, and external flows towards other countries. In the last century southern Europe has mutated from a region of emigration, to an area of transit of migrants heading towards northern Europe. As from the 1980s Southern European countries have become a destination for immigrants preceeding from poorer southerner and eastern regions.[1]

The transition described above has been especially evident for EUMed countries, Greece, Spain and Italy, which are recently facing intense rates of migrant inflows (King et al. 2000; de Zulueta 2003; Schmoll et al. 2015). In order to explain this phenomenon we need to refer to what different authors have called the "Mediterranean model of migration", in which seasonal migrant workers have replaced local workers in many low-paid manual and unskilled jobs, for example as domestic workers, porters, agricultural laborers, or unskilled manufacturing (Baldwin-Edwards and Arango 1999; King et al. 2000).

The emergence of a such model needs to be investigated in light of the changes that have characterized production systems for the Mediterranean countries since the 1970s, in the transition from the fordist model of production to the post-fordist one,

---

[1]The first laws that concerned immigration were developed in 1985 in Spain, 1986 in Italy and 1991 in Greece. The case also applies to Ireland and Scotland, which have also turned into countries of immigration in recent decades (Jentsch and Simard 2009).

in which labour flexibility was to become more relevant. In the Mediterranean labour market. This has meant a push towards low wages, informal and precarious work, and growing degrees of informality and flexibility especially where labour is mostly physical and unskilled.

As Mingione (1995) highlights for southern Italy, the local labour market was characterized by a dualistic nature, with protected employment conditions in the public sector, and informal, precarious, and underpaid employment conditions for unskilled work in the private sector (i.e. construction, domestic work, and agriculture). Moreover, the entrepreneurial fabric characterized by small and medium family businesses contributed to encouraging the demand for seasonal, cheap and mostly unskilled labour.

In such a framework, the interconnections between internal and international migrations have characterized the development of EUMed countries. As local labour markets are affected by out migration the labour supply is substituted through new flows of workers internally and transnationally. In the aftermath of the World War II mass-migration from Southern to Northern European countries resulted in a "reserve army" for the mining and industrial sectors of Northern Europe (Castles and Kosack 1973; Piore 1979; Pugliese 1993; for a discussion see King 2000). Most of these workers, together with their families, originated from poor rural areas, often from Southern regions of EUMed countries. In the 1960s, the modernization processes that took place in EUMed increased the levels of income and livelihoods reducing the development gap between Northern and Southern Europe, which contributed to migratory patterns. Southern European countries developed into industrialized areas, with increasing labour demands. This created new patterns of internal migration similar to the EU-wide pattern described above. In Italy and Spain grants began moving from the rural South to the Industrial North. According to King (2000: 10):

> this movement involved the transfer of mostly unskilled and poorly educated workers from low-productivity agricultural jobs to high-productivity industry and services and this migration was fed by a buoyant rate of demographic growth which sustained 'unlimited supplies of labour' (Lewis 1958).

The struggles of workers in the 1970s broke this system as they refused to be "squeezed" in this way. Meanwhile, the rural exodus dried up, also due to falling birth rates, in particular in the South and in general in rural areas. Throughout these recent decades, while the agricultural sector decreased consistently in its relative importance within the EU economy, it nonetheless continued to demand a seasonal, flexible and precarious workforce, as agriculture has become decreasingly attractive to the local populations.

In the EUMed rural settings affected simultaneously by an economic and demographic crisis the relationships between patterns of emigration and those of immigration became critical. As Table 3.1 and Fig. 3.1 show the rate of population aged over 65 years has increased in the last decades: in 2017 the 65 years old represent 23% of total population in Italy, 20.4% in Greece and 19.4% in Spain. In the 1960s this percentage was less than 10%. In the same period, the rural population decreased and the international migration stock strongly increased.

Table 3.1 Recent demographic trends in EUMed countries

		Spain	Italy	Greece
Rural population				
2017	%	20	29,9	21,3
	Amount	9.277.148	18.078.231	2.289.387
2000	%	23,7	32,6	27,1
	Amount	9.630.000	18.664.484	2.948.257
1990	%	24,7	33,3	28,5
	Amount	9.580.406	18.872.760	2.909.451
1960	%	43,4	40,6	44
	Amount	13.227.520	20.400.656	3.671.291
International migration stock				
2015	%	12,7	9,7	11,3
	Amount	5.852.953	5.788.875	1.242.514
2000	%	4,1	3,7	10,1
	Amount	1.657.285	2.121.688	1.111.665
1990	%	2,1	2,5	6,1
	Amount	821.605	1.428.219	618.139
1960	Amount	210.897	459.553	52.495
Population over 65				
2017	%	19,4	23	20,4
	Amount	9.051.923	13.939.689	2.194.721
2000	%	16,7	18,1	16,4
	Amount	6.776.002	10.333.787	1.770.442
1990	%	13,4	14,8	13,5
	Amount	5.069.944	8.425.807	1.381.186
1960	%	8,2	9,5	7,0
	Amount	2.494.868	4.766.245	587.257

Source: Worldbank – https://data.worldbank.org/indicator/

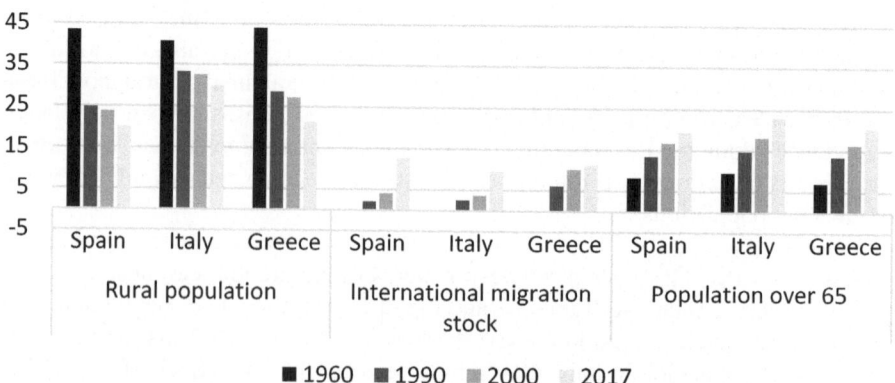

Fig. 3.1 Recent demographic trends in EUMed countries (%). (Source: Our re-elaboration on Worldbank data, https://data.worldbank.org/indicator)

Rising educational and living standards, the decreasing interest in manual work vis-à-vis other employment opportunities, and changes in the social and economic system (with the growing rates of local women in workforce, for example) has caused new needs for care workers in the private sector. The functioning of the Mediterranean model of welfare is characterized by a mix of state, market and family, with poor levels of public services, and the centrality of the family, and of women in particular, in the provision of care for the elderly and children (Ferrera 1996; Martin 1996; Pugliese 2000; Naldini 2003). As ironically noted by some researchers (Salazar Parreñas 2001; Ambrosini 2013), the growing amount of working women creates a new labour market for other women, usually immigrants who replace them in domestic labour, including in rural communities, where foregin caretakers and domestic workers upport local households and communities.

Today, the southern labour market still remains dualistic, with protected jobs in the public sector and informal and unstable, underpaid and unskilled work in most private sector employment (construction, tourism, care-giving, and agriculture).

Since the 1990s, the restriction of public spending reduced opportunities in public sector employment. These cuts were evident especially in southern regions, where public personnel typically exceed the real work needs. Emigration became more intense, while the private sector continued to demand for manual, unskilled and temporary workers in agricultural, but also in domestic work, private care, tourism and catering, construction. Immigrants eventually came to fill the gaps left by declining local populations and changing labour patterns, for cheap and flexible labour (Mingione and Quassoli 2000; King 2000; Osti and Ventura 2012).

Given the generalized informalization and downward trend in wages (especially for manual work with low productivity in services, agriculture and construction) over several decades, the EUMed has seen many rural areas transition from net-emigration to net-immigration regions.

According to King (2000: 14) this model of migration in the EUMed today is characterized by these features:

1. A kaleidoscope of different nationalities: while in the past migrations saw the predominance of specific groups depending on the destination countries. The multiplicity and heterogeneity of migrant nationalities today is stratified in a local labour market segmented by ethnic groups (called the "ethnicization" of the labour market, e.g., specific jobs are carried out mostly by workers coming from the same countries of origin.
2. Highly gender-specific flows from different countries: while previous migration involved male workers, recent migrations are mixed and segmented in relation to the kind of job. Many authors have reported over this territorial reconfiguration of ethnic as well as gender specialisation, with distinct communities occupying specific ecological and productive rural niches (Schrover et al. 2007; Bell and Osti 2010; Ambrosini 2013).
3. An increasing proportion of urban, educated persons: while in the sixties migrants were typically poor, less-educated and less-qualified, since the eighties the regional and social origins of migrants have become varied and stratified.

Migrants come from both rural and urban regions, with differentiated ages and higher levels of education. The work they are undertaken is often under-qualified compared to the work in the country of origin or to their level of education. Agriculture often remains often a refuge sector for the poorest and least educated immigrants

4. High levels of "clandestinity and irregularity", whereby the labour market has "become a global 'industry' with its own economic logic and market characteristics, with high and increasing levels of illegality as well as of informality (Castles and Miller 1993; King 2000: 13).

3.4   Migrant Workers in EU Mediterranean Agriculture: Some Common Features

Farm labour in the EU is undergoing a long-term decline; in 2017 the EU agricultural sector emploied about 8.9 million people, almost a fourth less (-25.5%) than it did in 2005. This decline is mostly due to two complementary and opposite trends, rising agricultural productivity due to mechanisation has reduced the need for labour in some areas, while land abandonment has decreased options for labour in others. While family labour in agriculture has been constantly reducing, opportunities for hired labour have proportionally grown; during recent years (2011–2017), hired labour has increased by 1% per year on average (Natale et al. 2019).

This phenomenon is common through the globe, as the International Labour Organization (ILO) reported at 40% the rate of hired or wage workers over employees in agriculture in the world in early 2000 (Pigot 2003). The reconfiguration of agriculture have thus carried relevant implications on the agricultural workforce, which in recent decades has shifted from mostly a family one, to hired, salaried workers, which have eventually shifted from a local to foreign ones. As a result agricultural activities today in Europe are increasingly carried out by immigrants.

In these areas industrial and commercial agriculture request wage workforce for harvesting in fruit and horticulture, as well as in labour-intensive crop and livestock farms. Due to its characteristic features (seasonality, high-intensity, manuality, low-skilled, mobility), agriculture in Mediterranean Europe represents an entry sector into the local labour market. The little interest of native workers provides little room for competition, and the greater tolerance for informal contractual conditions provide further elements for newcomers to seek opportunities in agriculture. Today more than a third of the officially employed agricultural workforce in Greece, Spain and Italy is of foreign origin (Collantes et al. 2014; Kasimis et al. 2010; Caruso and Corrado 2015; Corrado 2017; Papadopoulos and Fratsea 2017; Nori 2017).

Estimating the weight of migrant labour in agriculture is though very complex due to the high level of informality, and the heterogeneity in the quality and type of migration statistics available. In 2013, CREA estimated that 37% of agricultural

Table 3.2 % immigrant workers in EUMed countries

% Employed persons with foreign citizenship on total employed persons[a]				
	EU-28	Spain	Italy	Greece
2017	8,0	11,1	10,6 (16,6 in agriculture[b])	5,7
2008	6,6	13,9	7,4	6,6
% Immigrants on wage labourforce in agriculture, (estimates)[c]				
		Spain	Italy	Greece
2013		24	37	>50
2008		19,1	19,4	17

Source:
[a]Eurostat – https://ec.europa.eu/eurostat/web/migrant-integration/data/database
[b]Ministero del Lavoro (2018), Eighth Annual Report on Foreigners in the Italian Labour market,
[c]Data sourced from Caruso and Corrado (2015). For Spain: OMT, Observatorio Mercado del Trabajo. Madrid; OPI, (2014). Observatorio Permanente de la Immigraciòn. Madrid; For Italy: CREA (2015); For Greece: HELSTAT, (2013). Hellenic Statistical Office. Athens

salaried workforce in Italy was of immigrant origin. Using the ISTAT sample survey on the labour force, in 2017 the Italian Ministry of Labour estimates the proportion of immigrants to be 16%. The Labour Market Observatory (OMT) in Spain reports this rate to be 24% in 2013, while the Hellenic Statistical Office (ELSTAT) indicates that more than half of the salaried workers in agriculture are of foreign origin (Table 3.2).

In the last decade, after the financial crisis in 2008, intense immigration has represented a key factor of resilience for the agricultural sector and the rural world in these countries, as it has enabled many farms, rural villages and agriculture enterprises to remain alive and productive throughout difficult times (Kasimis 2010; Sampedro 2013; Collantes et al. 2014; Caruso and Corrado 2015; Schmoll et al. 2015; Nori 2017).

Conversely, and perhaps ironically given the worsening labour conditions, the rural world has also played a relevant role in enhancing the resilience of immigrant communities to the recent economic recession, following the worsening of living conditions in their home regions or of employment opportunities in destination urban areas. The majority of immigrant workers have recently found employment in EU rural areas rather than in the traditional urban migration centers, with important proportions finding work in even more peripheral parts of rural areas of the EU, for example in the mountainous or island rural communities (Jentsch and Simard 2009; Kasimis and Zografakis 2012; Colucci and Gallo 2015).

This may be because Mediterranean agriculture is a labour niche providing immigrants with better access to affordable basic resources like food and shelter and to more accessible employment opportunities. Mediterranean agriculture allows migrants to find options for employment and income, while cutting down on expenses as they build savings or send remittances.

These areas also offer degrees of non-visibility and informality that help accommodate their needs for decreased surveillance by the state, while creating as well the enabling environment for illegal practices and harsh exploitation, as it will be assessed.

Workforce in EUMed agriculture is also characterized by strategies of circular mobility, with workers moving between the different production areas according to seasonal peaks for labour, while moving back home when demand is low. The majority of workers in circular migration patterns come from Eastern Europe and the Balkans, as they face less trouble when it comes to documentation and visa permits and also because transportation costs are lower. The different capacities, opportunities and rights characterizing citizens proceeding from different regions depend on the different legislative and political agreements their countries of origin have stipulated with the EU or with its member states (as reported in Appendix).

As rights and duties change from a region to another, this provides room for employers and farmers to play with these varying and shifting conditions. Strategies of substitution of migrant groups according to criteria such as nationality, gender and legal status have continuously availed the agricultural sector of a cheap and vulnerable workforce. Literature reports about cases whereby the replacement of some groups with others are linked with attempts to restrict collective action and negotiation power of the former (Lindner and Kathmann 2014; Lenoël 2014; Sampedro and Camarero 2015; Caruso 2016b; Corrado 2017).

3.5 Rights at Stake

Several studies and reports point out to the largely informal contractual relationships and the precarious living and working conditions that characterize EUMed rural areas and that involve illegal hiring, low wages, high workloads, and suboptimal working and living conditions (Baldwin-Edwards and Arango 1999; King et al. 2000; OECD 2012; Colloca and Corrado 2013; Gertel and Sippel 2014; Corrado et al. 2018). Social researchers have been increasingly working to document the systematic denial of the basic rights of women and men which underpin most of EUMed's agricultural system. These conditions imply high social costs for the whole society, undermining the social acceptance of certain agricultural systems as well as their sustainability (Ortiz-Miranda et al. 2013; Gertel and Sippel 2014; Corrado et al. 2016, 2018; Nori 2017; Oxfam 2018). The exploitative nature of the restructuring of agricultural labour markets in the EUMed remains, however, still largely invisible to consumers and to citizens.

Reports are that several immigrants in EUMed agriculture often work for 10–12 h a day, for a wage that is considerably below the legal minimum one. Amongst the many difficulties and problems experienced by precarious immigrant workers in rural areas of the EUMed, a main one is linked to the absence of inclusive policies. In most agricultural intensive areas, immigrant workers live in rural slums, isolated and socially segregated from local populations. "Ghetto economies" is the name given to these highly exploitative situations where migrant workers are forced to live in appalling makeshift living conditions (Mangano 2014). The high concentration of precarious migrants in slums largely impacts on their capacity to integrate into local societies, and contributes generating racialized tensions and violences (i.e. the infamous episodes of El Ejido in Spain in 2001, Castelvolturno in 2008 and Rosarno in 2010 in Italy, and Manolada in Greece in 2013 amongst many others). According to Corrado et al. (2018: 24):

> The recurrent episodes of violence or racism by local populations towards seasonal workers—highlight the ambiguous coexistence of economic demand for migrant labour in the fields and social hostility to their presence in the streets. Wage gaps, precariousness, marginalisation and extreme flexibility are recurring elements in all Mediterranean countries but are play out differently in the different context.

For those residing far from the working fields, rural mobility and transportation represents another form of exploitative condition; costly and unsafe transportation services have resulted in numerous fatal accidents for the workers. Altogether, housing and transportation have caused dozens of deaths of immigrant workers in southern Italy in recent decades.

Other situations exist whereby farmers provide housing on the farms, subtracting these and other living costs from the wage of workers (e.g. in livestock farms). These immigrant workers living in isolated farms in the countryside suffer often loneliness and isolation.

Unhealthy working conditions might add up to physical exhaustion from excessive hours. Agricultural workers engaged in farms, barns and greenhouses are exposed to harmful chemicals, pesticides, and fertilizers. Illnesses and injuries associated include heat and water stress from working in the hot sun, exposure to cold and wet conditions during winter times, long-term chemical exposure under greenhouse plastic that emits harmful fumes.

Recruitment and engagement in the labour market are also domains where immigrants might face high costs and risks, as the official channels of employment are oftentimes ineffective. In Italy this has generated the phenomenon of 'caporalato'.

Box: The Caporalato System

The "caporalato" is a labour contracting system in the informal sector in Italy. An intermediary coordinating the relationship between migrants and farm owners, is part of the caporalato system. This system is traditionally characterized in harvesting of tomatos and citrus fruits in the South of Italy (Colloca and Corrado 2013; Pugliese 2012; Perrotta and Sacchetto 2013; Perrotta 2015; Corrado et al. 2018) and has become widespread as well in the wine sector in Northern Italy (Mangano 2013; Donatiello and Moiso 2017).

An emblematic case is the tomato harvesting in Apulia (Perrotta 2016). The seasonal pickers live in informal slums in rural areas, often far from villages, and are typically paid on a per-piece basis (i.e. 3–4 € for every 300 kg box of picked tomatoes).

(continued)

The workers are prevalently from sub-Saharan Africa and are recruited by different caporalato systems, organized often by other immigrants. The *caporale* (illegal or informal labour contractor) intermediates between the workers and the farmers, or may organize a work-team with relatives, friends and members of the same national community. According to the farmers, the caporalato system, also provides the housing and transportation for workers (often paid on the workers' dairy salary), which creates opportunities for further harassment, considering the absence of other means of transport to reach the land.

According to Perrotta (2016), this system in the tomato sector is the result of different factors including: (i) a lack of a public recruitment system; (ii) the inefficiency of the supply chain organizations and the small size of tomato processors; (iii) the pressure from retailers; (iv) the absence of investment in the mechanization of harvesting operations.

A Law approved in 2016 has defined the "caporalato" as a crime and has forged legal instruments accordingly.

These critical factors tend though to remain largely unaddressed because new supplies of precarious immigrant workers are constantly available. Since seasonal immigrants coming to the EUMed from Africa, Asia and Eastern Europe face oftentimes problems with their legal status and related vulnerabilities, they can easily become victims of blackmailing and several other forms of exploitation. The replacement and circulation of workers remains high, both because there is a large "reserve army" and because agriculture remains a "buffer" sector, where immigrants take refuge when they lose a safer job.

Despite these negative aspects, agriculture continues to be seen as an area of opportunity for immigrants to enter the labour market. After this broad introduction on the conditions of immigrant labour in the agriculture sector in the EUMed, we now turn to more specific analyses in the three EUMed countries: Greece, Spain and Italy.

3.6   Country Cases: Greece, Spain and Italy

3.6.1   Greece

The agricultural sector in Greece is considered a source of supplementary income for Greek workers and represents a buffer for local rural economies where non-farm employment is unstable. Agriculture still provides vital support for a significant number of rural areas in Greece (Kasimis 2010). Changes in the rural economy over the past three decades, however, have triggered a restructuring of rural Greek society, whereby the massive rural exodus of the 1960's and the expansion of other, non-agricultural activities have caused labour shortages that have not been filled by the local population in rural Greece (Kasimis et al. 2010; Kasimis 2010:261).[2]

New transnational immigrants have thus come to fill this gap by replacing the vanishing local population. In Greece, immigrants cover seasonal labour needs, contribute to increasing agricultural production, and help to keep wages and agricultural product prices low (Lianos et al. 1996; Vaiou and Hadjimichalis 1997). The geographical proximity of Albania to Greece, led to the development of a circular migration and recruitment system of Albanian labourers in Greece (Labrianidis and Sykas 2009) in the late 1990s. The studies carried out by Kasimis et al. (2003, 2010) in the early 2000s offer interesting insights into the economic and social implications of the settlement and employment of migrant labour in rural areas.

Immigrants played diverse roles in the survival of Greek agriculture and the maintenance of the rural social fibre. Their versatile skills and geographical mobility over multiple seasons provided a highly flexible labour force supporting the survival, expansion and modernisation of farms. The availability of a migrant workforce helps counterbalance the outmigration of the local workforce, keeping production costs low and thus enabling the agricultural activities to continue even in marginal settings, such as in the mountainous areas and the islands that cover a large part of the Greek territory (Kasimis and Papadopoulos 2005; Kasimis et al. 2010; Ragkos et al. 2016). Some of the contributions of immigrants to the social fibre of rural Greek communities include the revival of traditions through the use of traditional materials, and the demographic renewal of marginal rural areas through introducing higher fertility rates and through intermarriage with local residents (Kasimis 2009; Kasimis et al. 2010; Gallo and Rioja 2016).

Initial inflows of Albanians have been followed by those of Bulgarians and Romanians, and more recently by refugees who have reached Greece from neighbouring war-torn regions. Particularly in the aftermath of the recent economic crises, these more recent groups of immigrants have increased their geographical mobility between different rural areas in response to their precarious position, labour insecurity, and low socio-economic status (Papadopoulos 2012; Papadopoulos and Fratsea 2016). Several bilateral agreements have facilitated a process of seasonal/circular movement with Albania, Bulgaria, and Egypt (Triandafyllidou 2013) coming to rural Greece to work. The introduction of a three-month visa for Albanian nationals has made it easier for them to take up seasonal work in peak seasons, but some may end up working without legal status. Every 2 years a joint ministerial decision sets the maximum amount of positions for seasonal employment by region and sector. Non-EU citizens can be admitted to work for a maximum of 6 months through an "invitation" or "call" system (*metaklisi*), for which it is however difficult to apply for (Corrado et al. 2018).

---

[2]Relevant sources for data on migrants' presence in the Greek rural setting include the Hellenic Statistical Office (ELSTAT), the Greek Ministry of Migration Policy as well as the Hellenic Foundation or European and Foreign Policy (ELIAMEP, www.eliamep.gr).

In April 2016, the law was amended (Art. 13a Law 4251/2014) so that farmers/ employers in regions where seasonal working positions exist, and have already been approved, may recruit irregular third-country nationals or asylum seekers already resident in Greece. This thus has enabled the regular employment of irregular migrants by providing them with a temporary, 6-month permit (Papadopoulos and Fratsea 2017). This device is however criticised as it generates even further dependence on employers, since when that period expires, the suspension is lifted and the rights of workers are lost (Corrado et al. 2018).

Box: Strawberry Production in Manolada

As Corrado et al. (2018) note strawberry production in Manolada is characterized by a significant expansion and greenhouse systems. Several factors have contributed to the growth in strawberry cultivation: the replacement of fresh strawberry plants with frozen ones, thus allowing for a longer harvest period and better organoleptic characteristics; the establishment of an export-oriented cooperative; and, finally, the availability of cheap and flexible migrant labour (Papadopoulos and Fratsea 2016).

While the number of Albanian, Bulgarian and Romanian workers in Manolada has progressively decreased, the amount of Bangladeshis has increased. Predominantly single males, with very low educational profiles, Bangladeshi migrants live in collective houses or in makeshift tents. Most of them are without legal status, which significantly increases their vulnerability to exploitation.

In 2013, 150 Bangladeshi workers went on strike in Manolada to claim unpaid wages. The companies's armed guards fired on the protesting workers, severely injuring 30 of them. The case was brought before the European Court of Human Rights (ECHR), which in 2017 judged on that Bangladeshi migrant workers' conditions were those of forced labour and human trafficking (Corrado et al. 2018).

Greek was accused of violating article 4 of the European Convention on Human Rights, but no significant measures have been implemented to prevent and address this form of exploitation so far. That same year, Open Society foundation in partnership with a Greek NGO Generation 2.0 initiated a paralegal project with a view to build a community infrastructure in Manolada, by providing, amongst other measures, mobile legal clinics to address legal needs and rights issues of migrant workers (Corrado et al. 2018).

3.6.2   Spain

The recent growth of immigrant workers in rural areas started in the 1980s and 1990s when Spanish agriculture became more intensive. In the early 2000s, immigration increased further, spurred by the economic crisis in 2008 that collapsed opportunities for employment in other sectors (for example the construction sector). Today, immigrants contribute between one quarter and one third of the salaried agricultural workforce in Spain, these include many precarious workers. These workers contribute significantly to enhancing productivity in the agricultural sector as well as to reverse depopulation trends, ensuring the survival of local economies in many rural areas (Collantes et al. 2014).[3]

While rural immigrants mostly contribute to agriculture production in the intensive systems that characterise the Southern regions of the country, they also represent a relevant and increasingly appreciated resource to maintain populations in rural villages, especially in remote areas (e.g. the Nuevos Senderos program[4]) (on the Spanish case see Hoggart and Mendoza 1999; Esparcia 2002; García Coll and Sánchez 2005; Solé 2010; Prieto and Papadodima 2010; Camarero et al. 2013; Collantes et al. 2014; Gallo and Rioja; 2016). The mobility of rural migrants is concentrated around three main hubs: (i) the neighbouring provinces of the eastern area: Murcia, Alicante, Albacete and Almería; (ii) between Barcelona and the Catalan provinces of Tarragona and Girona; (iii) between Valencia, Alicante, Castellón and Barcelona (Observatorio de las Ocupaciones 2014; Viruela and Torres Pérez 2015; Caruso and Corrado 2015).

The rural agricultural labour market in Spain is complex in terms of spatial and temporal organization. The local labour market is segmented, with different migrant groups competing with each other and with local workers in the different agricultural areas in different work-intensive periods, from the winter harvest of olives and the spring harvest of strawberries, to other vegetables and fruits, but this also applies to variation in labour requirements in livestock breeding systems. Circular mobility schemes amongst different agricultural areas mostly include provinces of Andalucía and Catalonia.

Traditionally, many low-wage agricultural workers came from Latin America and the Middle East and North Africa, specifically Morocco (Checa 2001; Hellio 2016). In recent years, there has been an intensifying presence of immigrants from sub-Saharan Africa and from Eastern Europe as well as Asian countries in rural areas, as a result of the economic crisis discussed earlier which made urban employment options less available in migrants' home regions.

Spain is an interesting case because the recruitment of seasonal migrant workers is stimulated and managed by national policies, like those in non-European countries

---

[3]Relevant sources for data on migrants' presence in the Spanish rural setting includes the Instituto Nacional Estadistico (INE) with its Encuestas de Variaciones Residenciales, the Observatorio Mercado Trabajo (OMT), and the Observatorio Permanente de la Inmigración (OPI).

[4]http://cepaim.org/que-hacemos-convivencia-social/desarrollo-rural/nuevos-senderos-empleo-rural/

including the US, Canada, and Australia (Martin 2016). In Spain, the seasonal workers are recruited directly in their countries of origin through a *"contratación en origen"* and a quota system known as the *Régimen General* and the FNAAC (Framework National Agreement on Seasonal Workers for Agriculture Campaigns). On the surface, this model seems to be a flexible and consensual system matching the demand and the offer of labour through public governance that includes public institutions, private actors and business associations (as intermediaries). However, similar to other national programs based on quota policies restricting the amount and length of stay of internationally recruited workers, this model exposes immigrant labourers to high levels of exploitation and abuse: the immigrants became strongly dependent upon intermediaries and employers.

As summarized by Corrado et al. (2018: 25):

> the private intermediation of farmworkers is ensured by Temporary Employment Agencies (*Empresas de trabajo temporal*, ETT) regulated by Law No. 14/1994), which largely control contracts in areas like Valencia or Murcia. During a harvesting season, workers can have several contracts. ETTs move workers across regions, provinces, or countries and play a fundamental role in the ethnic segmentation, replacement, and rotation of the labour force to ensure flexibility but also causing job insecurity.

Many researches investigate this system of exploitation, including Moreno (2009 2012), Martin (2016), Pumares and Jolivet (2014), Reigada (2016), Gadea et al. (2016), Avallone and Ramirez-Melgarejo (2017), Corrado (2017).

Box: Immigrants Substitution in Spanish Agriculture: The Case of Huelva

Following Lindner and Kathmann (2014: 127) "over the last two decades, Huelva has emerged as Europe's most developed strawberry-growing region, annually producing 260,000 tons of strawberries — roughly 35 per cent of the entire European production. If the autumnal work of planting requires 1.000 workers, at least 60.000 labourers are needed for harvesting". By the end of the 1980s foreign workforce started substituting the Spanish one, as agricultural work was associated to a low social status. It was then mostly Moroccans residing in Spain that came to contribute their labour to this sector, a less vindictive and cheaper workforce compared to the local one. In the same period Spanish women started to engage in the previously men-dominated packaging industry.

As a result of the riots in El Ejido (Almeria province) in 2000, which involved Moroccan agricultural labourers demanding for better working and living conditions, agricultural entrepreneurs in most of Southern Spain started looking for 'less conflictive' and 'more adaptable' labour migrants. In the early 2000s workers from some Latin American countries (such as Ecuador) but mostly from Eastern Europe (mostly Poland and Romania) increasingly entered the Spanish agricultural labour markets, substituting the Moroccan

(continued)

workforce. As to Lindner and Kathmann (2014: 128) "for several years, in the spring months, bus caravans from Poland and Romania brought thousands of women to Huelva to meet the demand for labour; 7.000 Polish women in 2002 were taken to Huelva. The least productive workers were sent back after 15 days of trial or when they were no longer needed for the harvest peaks".

Following the enlargement of the EU in 2007, and with the acquisition of better rights, Eastern European women were less bound to seasonal work in agriculture. Their substitution begun thus through these seasonal labour migration schemes denominated *contraction en origen* whereby female workers were directly recruited in the rural areas of Morocco, preferably with young children – as a critical incentive to return home by the end of the temporary contract (Hellio 2016; Caruso 2016a). The interplay between Romanian and Moroccan female workers that has characterised recent years, until the more recent entry of irregular African migrants into the Spanish agricultural labour market, has provided further availability of a cheap workforce.

3.6.3 Italy

Although it is not easy to ascertain these data according to CREA (2017) the amount for foreign workers in Italian agriculture was around 405.000 in 2015, +5% compared to 2014. 27% of these are women, and over 40% of the immigrant workers are found in Northern Italy. As to Table 3.3 EU citizens remain the most numerous migrant workers in Italy (over 211.000 people), with a modest recent increase compared to those who are not citizens of the EU. In particular, the growing amount of Eastern European workers is due to the EU enlargement that has facilitated the entry of Romanian workers, whereas many refugees have become agricultural workers in the last 2 years.[5]

Table 3.3 Migrant workers in Italian agriculture by country of origin

Country of origin	2008	2016
Romania	77.250	112.289
India	9.867	26.900
Albania	17.018	24.870
Morocco	14.435	23.932
Poland	24.708	15.986
Bulgaria	14.482	12.036

Source: INPS data elaboration, 2016 in Corrado et al. 2018

---

[5]Relevant sources for data on migrants' presence in the Italian rural settings include Osservatorio Placido Rizzotto (OPR), Istituto Nazionale di Statistica (ISTAT), Istituto Nazionale Economia Agraria (INEA), Caritas.

Agricultural workers employed without an official contract are estimated at 430.000, of whom around 80% were foreign workers and about 100.000 were identified as being at a high risk of exploitation. Women comprise about 42% of farm workers without legal documentation and are usually over-represented in unpaid and seasonal work (Dines and Rigo 2015; Medu 2015; OPR 2018; Oxfam 2018).

There are many differences between the conditions for workers in the different Italian regions: in Southern Italy, the workers are employed in harvesting and the work is irregular, informal and seasonal. In the Northern regions, workers are employed in the intensive livestock breeding sector, and the labour is more continuous and durable. According to Corrado (2017), non-EU workers tend to be young men employed in low-skilled horticultural jobs requiring physical strength. But women from non-EU nations are also employed, especially in Southern Italy in packing and processing jobs. This is particularly true in Sicily, where there is a consistent amount of immigrants engaged in agriculture (about 47,000 workers according to CREA). Research has documented the exploitation of Romanian female workers in greenhouses where they are often victims of blackmail and violence (see e.g. Palumbo and Sciurba 2015; Piro and Sanò 2016; Sanò 2018).

In the seasonal work patterns in Southern Italy, especially in Calabria and Puglia, but also in Sicily and Basilicata, workers are underpaid and official contracts only record and report a part fo the story, in kinds of 'grey' arrangements. Even when the immigrants have signed a regular contract, the declared hours and working days are reduced by the employees, increasing the exploitation and precluding the possibility to accrue social rights and benefits.

The case of Southern Italy is emblematic due to its seasonal and progressively specialized agriculture. In Calabria, Sicily, Campania, Apulia, and Basilicata open-air or greenhouse seasonal productions of fruits and vegetable rely mainly on small and medium-sized farms; products are oriented to fresh consumption or processing and serve distant distribution and corporate retailers. Furthermore, in agriculture, as much as in the general economy, the labour market in Southern Italy is characterized by informal contractual relationships. The immigrants move from one place to another depending on the seasonality of harvesting tomatoes, oranges, lemons, grapes and other fruit and vegetables.

Much research has documented the exploitation of immigrant people, in particular the non-European workers (that are often without legal documentation and thus more easily blackmailed) through the caporalato system in which immigrants are informally hired for very little money or through other forms of exploitation (AA. VV. 2012; Colloca and Corrado 2013; Pugliese 2012; Perrotta and Sacchetto 2013; Perrotta 2015; Corrado et al. 2018).

Another way in which a precarious labour supply is organized for farmers is through temporary staffing agencies, or service agencies (particularly organizing the labour of Romanian workers) and through the mechanism of the so-called «landless cooperatives». This is a way to hire immigrant workers from Eastern Europe for the seasonal harvest. They are engaged as «worker members» of the cooperatives (that in Italy have a special and facilitated tax regime), thus employees manage to evade social security obligations, minimum legal wages, and labour laws regarding working hours. The research of Caruso (2016b) and Perrotta (2014) investigated the phenomenon of "landless cooperatives" in the South. Recently, Donatiello and Moiso (2017) have documented the same mechanisms in the harvest of grapes in Canelli (Piedmont in the North) for the production of D.O.C. (Protected Designation of Origin, certified through a quality control system) wine. The Macedonian migrants who have settled in Canelli for several years have a fairly stable status, and have organized a system of exploitation of other workers from Eastern Europe (in particular the Romanians) who are hired seasonally through landless cooperatives and are mostly precarious, exploited and underpaid.

The systems described (caporalato system, the landless cooperatives, and the staffing agencies) indicate important degrees of collusion amongst various agents of the agriculotural sector, as well as the complacency of a legal and policy framework that provide an enabling environment for grey practices at the expenses of immigrant workers.

This collusion between immigrant labour brokers and farmers is highlighted by Avallone (2016, 2017) in the case of the Piana del Sele in Campania, where immigrants work in greenhouses that produce pre-mixed salads for supermarkets. In this area, the "decreto flussi" – the Italian programme for the recruitment of foreign seasonal workers – has often been used by local farmers and illegal brokers to cheat migrants and the state alike. The local market is segmented by nationality and gender. Male workers from Morocco (the first to work in this area) are today stressed by competition with Romanian immigrants who accept lower salaries: the daily wage (7 h) varies from 27 to 33 Euros and the monthly salary ranges between 500 and 800 Euros (Avallone 2016).

Another field of research analyses the extensive production of canned tomatoes in the area of Vulture-Alto Bradano in Basilicata, which has employed many African people (Burkina Faso, Mali, Ghana, Tunisia, Morocco are the most frequent countries of workers' origin) (Perrotta 2016). The cases of lemon and orange harvests in Calabria illustrate some of the same patterns described above (see box).

> **Box: Picking Citrus Fruits in the Plains of Sibari and Rosarno, Calabria**
> According to Corrado (2017), in the Calabria region migrants are employed in the winter for harvesting lemons and oranges, with some local differences. This agri-food chain is exposed to major concentration in large scale distribution and to a high pressure from the global market that calls for a high volatility of orange prices, inefficient delivery, and flexible workers available to answer to the «just-in-time» demands of supermarkets (Garrapa 2016). In this sense, the work is very precarious, unstable and informal, involving different kind of vulnerable immigrant workers, who may or may not have legal status (as asylum seekers with temporary international protection, «rejected asylum seekers», or refugees, including minors, who have evaded identification

(continued)

procedures, Corrado and D'Agostino 2018). Many workers from African nations arrive in Calabria after the end of the tomato harvest season in Puglia and Basilicata, and after the end of the citrus season they move on, either to Campania or to Sicily, living in rural temporary camps that are real slums.

In the Plain of Sybaris, the harvest operations start in November and last until to July, with several different crop harvests (starting with the harvesting of olives, oranges and clementines, then pruning and grafting operations, and finally the peach and strawberry harvests). The immigrant workers come mainly from the Maghreb (Tunisia and Morocco) and from Eastern Europe (Romania, Bulgaria, and Ukraine), with differences in wages and working conditions between different groups of immigrants, as well as between them and local workers, demonstrating race-gender segregation in agriculture:

> The African workers receive, on average, 20–25 euros per day, [and] those from Eastern Europe. . . earn up to 35 euros, while [local] workers receive around 40 euros. The workday is generally 10–15 hours; piece rate work, paying 5 euros per box, is also common. Informal brokers (caporali) can retain up to 10 euros for their role in intermediation, transport and the provision of basic goods such as water. Women from Eastern Europe are paid 1,50–2,50 euros per hour (while men receive up to 3 euros) and work around eight hours a day. Some employers prefer to hire women-only teams: Their wages are lower, they are more consistent in their work activity, and they achieve higher work and production rates. Women leaving their children at home are often selected, because after the harvest season they tend to return home. That said, women in the field are often victims of sexual abuse and violence (Corrado 2017: 6).

In Calabria, we find some of the most difficult working conditions in an area of intensive orange production: the Plain of Rosarno is known for the first riots of sub-Saharan African pickers of oranges in 2010 (with clashes with the local population), countered by the national government with the forced relocation to other regions of 1,500 immigrant strikers.

As Corrado (2017: 7) has explained well, "the cut in EU subsidies for fruit production, coupled with a decline in market price for low-quality oranges used by the juice industry – from 1400 lire per kilo in 1999 to 10–20 cents in 2010, to the current price of 5–6 cents – due to low-cost imports from Brazil and Spain, has further discouraged harvesting. Because of this crisis, many growers are now converting to higher-value produce, such as kiwi, and Eastern European women are largely employed to harvest this new product". Furthermore, in the last years, there is a process of replacing African workers with Eastern Europeans both because they are willing to accept lower wages, and because since 2009, changes in migration laws have made it more convenient to hire these workers to avoid legal problems, since they can have legal status as members of EU nations.

3.7 Conclusions: Dynamics, Trends and Roles for Immigrants in Rural Areas

The Mediterranean is traditionally a region shaped by and through migrations. Its European shores have long been an area of emigration and have become today the nodal point of migratory flows to Europe. However, the European shores of the Mediterranean are not just a transit area, but also an attractive stop for migrants due to the high demand for low-skilled labour in agriculture.

The demand for cheap labour is particularly high in the regions where agricultural labour is normally temporary and precarious, requiring workers to move according to seasonal agricultural demands. Depopulating rural areas in the EU have become a haven for precarious, migrant workforce, as access to food, accommodation and employment can be less of a barrier than in traditional urban centers. In southern Europe, the agricultural workforce today mainly originates from North Africa and Eastern Europe, although recent flows involve refugees from conflict-ravaged areas and poor economic conditions, such as from countries deeper into the Asian and African continents.

This comes, however, with a cost. Under growing pressure from large production and distribution systems, agricultural producers tend to reduce production costs by employing low-paid labour under exploitative conditions. This results in agricultural systems that systematically deny the rights of workers, women and men, especially immigrants, who enjoy fewer citizenship rights, reduced salary and exploitative labour conditions.

Many studies point out to the 'grey' contractual relationships and the precarious conditions that characterize these working environments and to the related extensive social costs. These in turn generate stress in relationships between foreigners and local populations, as well as between immigrants of different national origins, and undermine the overall sustainability of this sector. This is particularly ironic when these exploitative practices take place in a sector that enjoys consistent public support through the CAP and other policies.

Migration patterns have been changing rapidly during the recent financial crisis and are likely to continue increasing in coming years for several reasons including political, economic and climatic ones. These flows will continue providing an important 'reserve army' for a sector that maintains salaries and rights low through mechanisms of substitution and replacements.

Appendix: The Conditions of Migrant Workers

Migration strategies differ depending on the specific rights and capacities the diverse communities might enjoy in the EU setting, in terms of concern circular migration patterns, family re-unification options, development of transnational networks, access to different services and opportunities, opportunities for local investments and entrepreneurial initiatives, etc.... These patterns carry important consequences on the communities of origin as well in EUMed countries.

Several mobility rights and duties are related to the Schengen framework, while others relate to countries' bilateral agreements.

Rights	Recent EU citizens (ie. Romanians, Bulgarians)	UE enlargment areas[a] (ie. Macedonians, Albanians)	EU Neighbourhood (ie. Moroccans, Tunisians)
Access and mobility	Unconditional rights; free circulation	Need for a Visa and related negotiation – although this process is being simplified due to Visa liberalisation throughout the Schengen area (e.g. Albanians no need visa below 3 months staying).	Need for a Visa and related negotiations within bilateral agreements – Cooperation on simplifying the procedures for access for certain categories of people (including the possibility of issuing multiple-entry and longer-term visas, and waiving administration fees).
Residential (above 3 months)	Since 2013 EU citizens	Residential rights associated to a) work permit or b) attachment o resident family member with whom there is direct dependency relationship – to be negotiated within the framework of bilateral agreements.	Residential rights associated to: a) work permit or b) attachment o resident family member with whom there is direct dependency relationship – Cooperation on simplifying the procedures for legal stays for certain categories of people to be negotiated within the framework of countries' bilateral agreements.[b]
Labour market – including possibility to set up an enterprise	Since 2013 unconditional rights	Simplified procedure for some countries (e.g. no need for visa renewal after 3 initial months). Possibility to get long-term residency and related non-discriminatory rights after 5 years of demonstrated legal presence in the country.[c]	Need to continuously renew residential visa. Possibility to get long-term residency and related non-discriminatory rights after 5 years of demonstrated legal presence in the country.[c]

(continued)

Rights	Recent EU citizens (ie. Romanians, Bulgarians)	UE enlargment areas[a] (ie. Macedonians, Albanians)	EU Neighbourhood (ie. Moroccans, Tunisians)
Family reunification	Unconditional rights	Possibility to invite family members after at least 1 year of legal residence subject to bilateral restrictions (related to (a) accommodation, (b) health insurance, (c) financial capacities). Process simplified within the ongoing Visa liberalisation negotiations.	Possibility to invite family members after at least 1 year of legal residence subject to bilateral restrictions (related to (a) accommodation, (b) health insurance, (c) financial capacities).

Source: Our elaboration, thanks to G. Renaudiere
[a]As part of the Stabilisation and Association Process (SAP) in 1999 and the EU Enlargement policy
[b]EU-Morocco mobility partnership (and the 9 participating Member States: the Kingdom of Belgium, the French Republic, the Federal Republic of Germany, the Italian Republic, the Kingdom of the Netherlands, the Portuguese Republic, the Kingdom of Spain, the Kingdom of Sweden, and the United Kingdom), June 2013 (http://ec.europa.eu/dgs/home-affairs/what-is-new/news/news/2013/docs/20130607_declaration_conjointe-maroc_eu_version_3_6_13_en.pdf)
[c]Nationals of these countries, who are working legally in the European Union, are entitled to the same working conditions as the nationals

References

Ambrosini, M. (2013). *Irregular migration and invisible welfare*. Houndmills: Palgrave Macmillan.
Arrighi, G. (1985). *Semiperipheral development: The politics of Southern Europe in the twentieth century (explorations in the world economy)*. Beverly Hills: Sage.
Avallone, G. (2016). The land of informal intermediation. The social regulation of migrant agricultural labour in the Piana del Sele, Italy. In A. Corrado, C. de Castro, & D. Perrotta (Eds.), *Migration and agriculture: Mobility and change in the Mediterranean area*. London: Routledge.
Avallone, G. (2017). *Sfruttamento e resistenze: Migrazioni e agricoltura in Europa, Italia, Piana del Sele*. Verona: Ombre Corte.
Avallone, G., & Ramirez-Melgarejo, A. (2017). Trabajo vivo, tecnología y agricultura en el sur de Europa. Una comparación entre la Piana del Sele en Salerno (Italia) y la Vega Alta del Segura en Murcia (España). *Ager. Revista de Estudios sobre Despoblación y Desarrollo Rural, 23*, 131–161. https://doi.org/10.4422/ager.2017.06
AA.VV. (2012). Brigate di solidarietà attiva). *Sulla pelle viva. Nardò: la lotta autorganizzata dei braccianti agricoli*. Roma: Derive&Approdi.
Baldwin-Edwards, M., & Arango, J. (Eds.). (1999). *Immigrants and the informal economy in Southern Europe*. Portland: Frank Cass.

Bell, M. M., & Osti, G. (2010). Mobilities and ruralities: An introduction. *Sociologia Ruralis*. Special Issue on Mobilities and Ruralities, *50*(3), 199–204. https://doi.org/10.1111/j.1467-9523. 2010.00518.x.

Camarero, L., Sampedro, R., & Oliva, J. (2013). Trayectorias ocupacionales y residenciales de los inmigrantes extranjeros en las áreas rurales españolas. *Sociología del Trabajo, 77*, 69–91.

Caruso, F. (2016a). Fragole amare: lo sfruttamento del bracciantato migrante nella provincia di Huelva. In Osservatorio Placido Rizzotto (Ed.), *Agromafie e Caporalato. Terzo rapporto*. Roma: Ediesse.

Caruso, F. (2016b). Dal caporalato alle agenzie di lavoro temporaneo: i braccianti rumeni nell'agricoltura mediterranea. *Mondi Migranti, 1*(3), 51–64. https://doi.org/10.3280/MM2016-003004.

Caruso, F., & Corrado, A. (2015). Migrazioni e lavoro agricolo: un confronto tra Italia e Spagna in tempi di crisi. In M. Colucci & S. Gallo (Eds.), *Tempo di cambiare. Rapporto 2015 sulle migrazioni interne in Italia*. Roma: Donzelli.

Castles, S., & Kosack, G. (1973). *Immigrant workers and class structure in Western Europe*. Oxford: Oxford University Press.

Castles, S., & Miller, M. J. (1993). *The age of migration: International population movements in the modern world*. London: Macmillan.

Checa, F. (2001). *El Ejido: la ciudad cortijo. Claves socioeconómicas del conflicto étnico*. Barcelona: Icaria Editorial.

Collantes, F., Pinilla, V., Sàez, L. A., & Silvestre, J. (2014). Reducing depopulation in rural Spain. *Population, Space and Place, 20*, 606–621. https://doi.org/10.1002/psp.1797.

Colloca, C., & Corrado, A. (Eds.). (2013). *La globalizzazione delle campagne. Migranti e società rurali nel Sud Italia*. Milano: FrancoAngeli.

Colucci, M., & Gallo, S. (Eds.). (2015). *Tempo di cambiare. Rapporto 2015 sulle migrazioni interne in Italia*. Roma: Donzelli.

Corrado, A. (2015). Lavoro straniero e riorganizzazione dell'agricoltura familiare in Italia. *Agriregionieuropa, 43*, 23–27. https://agriregionieuropa.univpm.it/it/content/article/31/43/lavoro-straniero-e-riorganizzazione-dellagricoltura-familiare-italia.

Corrado, A. (2017). Migrant crop pickers in Italy and Spain. *E-paper, Heinrich Böll Stiftung Foundation*. Berlin. https://www.boell.de/sites/default/files/e-paper_migrant-crop-pickers-in-italy-and-spain.pdf

Corrado, A., & D'Agostino, M. (2018). Migrations in multiple crisis. New development patterns for rural and inner areas in Calabria (Italy). In K. Stefan, W. Tobias, & I. Jelen (Eds.), *Processes of immigration in rural Europe: The status Quo, implications and development strategies* (pp. 272–295). Cambridge: Cambridge Scholars Publishing.

Corrado, A., De Castro, C., & Perrotta, D. (Eds.). (2016). *Mobility and change in the Mediterranean area*. London/New York: Routledge.

Corrado, A., Palumbo L., Caruso F. S., lo Cascio M., Nori M., & Traindafyllidou, A. (2018). *Is Italian agriculture a 'Pull Factor' for irregular migration and, if so, why?* Open Society Foundations. https://www.opensocietyfoundations.org/sites/default/files/is-italian-agriculture-a-pull-factor-for-irregular-migration-20181205.pdf. Accessed 20 Jan 2019.

CREA. (2017). *Annuario dell'agricoltura italiana 2015*. Roma: Crea.

De Zulueta, T. (2003). *Migrants in irregular employment in the agricultural sector of Southern European countries. Report for the debate in the standing committee*. Bruxelles: Council of Europe.

Dines, N., & Rigo, E. (2015). Postcolonial citizenships between representation, borders and the 'refugeeization' of the workforce: Critical reflections on migrant agricultural labor in the Italian Mezzogiorno. In S. Ponzanesi & G. Colpani (Eds.), *Postcolonial transitions in Europe: Contexts, practices and politics* (pp. 151–172). London: Rowman and Littlefield.

Donatiello, D., & Moiso, V. (2017). Titolari e riservisti. L'inclusione differenziale di lavoratori immigrati nella viticultura del Sud Piemonte. *Meridiana, 89*, 185–210.

ELSTAT. (2013). *Hellenic statistical office*. Athens: Coincise Statistical Yearbook.

Esparcia, J. (2002). La creciente importancia de la inmigración en las zonas rurales de la comunidad valenciana. *Cuadernos de Geografía, 72*, 289–306.

Farinella, D., Nori, M., & Ragkos, A. (2017). Change in Euro-Mediterranean pastoralism: Which opportunities for rural development and generational renewal? In C. Porqueddu, A. Franca, G. Molle, G. Peratoner, & A. Hokings (Eds.), *Grassland reources for extensive farming systems in marginal lands: Major drivers and future scenarios* (Grassland Science in Europe) (Vol. 22, pp. 23–36). Wageningen: Wageningen Academic Publishers.

Ferrera, M. (1996). The 'southern model' of welfare in social Europe. *Journal of European Social Policy, 6*(1), 17–37. https://doi.org/10.1177/095892879600600102.

Gadea, E., Pedreño, A., & de Castro, C. (2016). Producing and mobilizing vulnerable workers. The agribusiness of the region of Murcia (Spain). In A. Corrado, C. De Castro, & D. Perrotta (Eds.), *Migration and agriculture. Mobility and change in the Mediterranean area*. London: Routledge.

Gallo, R. S, & Rioja, L. C. (2016). Inmigrantes, estrategias familiares y arraigo: las lecciones de la crisis de las areas Rurales. *Migraciones, 39*, 3–31. mig.i40y2016.00840services.

García Coll, A., & Sánchez, D. (2005). La población rural en Cataluña: entre el declive y la revitalización. *Cuadernos Geográficos, 36*(1), 387–407.

Garrapa, A. M. (2016). *Braccianti just in time. Raccoglitori stagionali a Rosarno e Valencia*. Firenze: La Casa Usher.

Gertel, J., & Sippel, R. S. (Eds.). (2014). *Seasonal workers in Mediterranean agriculture: The social costs of eating fresh*. London: Routledge.

Hellio, E. (2016). They know that you'll leave, like a dog moving onto the next bin: Undocumented male and seasonal contracted female workers in the agricultural labour market of Huelva, Spain. In A. Corrado, C. de Castro, & D. Perrotta (Eds.), *Migration and agriculture. Mobility and change in the Mediterranean area* (pp. 198–216). London: Routledge.

Hoggart, K., & Mendoza, C. (1999). African Immigrant Workers in Spanish Agriculture. *Sociologia Ruralis, 39*(4), 538–556. https://doi.org/10.1111/1467-9523.00123.

ILO. (2015, December 15). *ILO global estimates of migrant workers*. www.ilo.org/global/topics/labourmigration/publications/WCMS_436343/lang%2D%2Den/index.htm

Jentsch, B., & Simard, M. (Eds.). (2009). *International migration and rural areas. Cross-national comparative perspectives*. London: Ashgate.

Kasimis, C. (2009). From enthusiasm to perplexity and scepticism: international migrants in the rural regions of Greece and Southern Europe. In B. Jentsch & M. Simard (Eds.), *International migration and rural areas* (pp. 75–98). London: Ashgate.

Kasimis, C. (2010). Demographic trends in rural Europe and migration to rural areas. *AgriRegioniEuropa, 6*/21. https://agriregionieuropa.univpm.it/it/content/article/31/21/demographic-trends-rural-europe-and-international-migration-rural-areas

Kasimis, C., & Papadopoulos, A. G. (2005). The multifunctional role of migrants in the Greek countryside: Implications for the rural economy and society. *Journal of Ethnic and Migration Studies, 31*(1), 99–127. https://doi.org/10.1080/1369183042000305708.

Kasimis, C., & Zografakis, S. (2012). *Return to the land: Rural Greece as refuge to crisis*. XIII world congress on rural sociology. IRSA, Lisboa.

Kasimis, C., Papadopoulos, A. G., & Zacopoulou, E. (2003). Migrants in rural Greece. *Sociologia Ruralis, 43*(2), 167–184. https://doi.org/10.1111/1467-9523.00237.

Kasimis, C., Papadopoulos, A. G., & Pappas, C. (2010). Gaining from rural migrants: Migrant employment strategies and socio-economic implications for rural labour markets. *Sociologia Ruralis, 50*(3), 258–276. https://doi.org/10.1111/j.1467-9523.2010.00515.x.

King, R. (2000). Southern Europe in the changing global map of migration. In R. King, G. Lazaridis, & C. Tsardanidis (Eds.), *Eldorado or Fortress? Migration in Southern Europe* (pp. 3–26). London: Palgrave Macmillan.

King, R., Lazaridis, G., & Tsardanidis, C. (Eds.). (2000). *Eldorado or fortress? Migration in Southern Europe*. London: Palgrave Macmillan.

Labrianidis, L., & Sykas, T. (2009). Geographical proximity and immigrant labour in agriculture: Albanian immigrants in the Greek countryside. *Sociologia Ruralis, 49*(4), 394–414. https://doi.org/10.1111/j.1467-9523.2009.00494.x.

Lenoël, A. (2014). *Burden or empowerment? The impact of migration and remittances on women left behind in Morocco*. PhD, Social Policy, University of Bristol.

Lewis, W. A. (1958). Economic development with unlimited supplies of labour. In A. N. Agarwal & S. P. Singh (Eds.), *The economics of underdevelopment* (pp. 400–449). Oxford: Oxford University Press.

Lianos, T. P., Sarris, A. H., & Katseli, L. T. (1996). Illegal immigration and local labour markets: The case of Northern Greece. *International Migration, 34*(3), 449–484. https://doi.org/10.1111/j.1468-2435.1996.tb00537.x.

Lindner, K., & Kathmann, T. (2014). Mobility partnerships and circular migration: Managing seasonal migration to Spain. In J. Gertel & R. S. Sippel (Eds.), *Seasonal workers in Mediterranean agriculture: The social costs of eating fresh*. London: Routledge.

Mangano, A. (2013). Piemonte: la vendemmia della vergogna. *L'Espresso*, 3 December. http://espresso.repubblica.it/inchieste/2013/12/03/news/il-prestigioso-vino-piemontese-e-prodotto-come-a-rosarno-1.144081

Mangano, A. (2014). *Ghetto economy*. Edizioni: Createspace Independent Publishing.

Martin, C. (1996). Social welfare and the family in Southern Europe. *South European Society & Politics, 1*(3), 23–41. https://doi.org/10.1080/13608749638539481.

Martin, P. L. (2016). *Migrant workers in commercial agriculture*. Geneva: ILO. https://www.ilo.org/wcmsp5/groups/public/%2D%2D-ed_protect/%2D%2D-protrav/%2D%2D-migrant/documents/publication/wcms_538710.pdf

Medu. (2015). *Terraingiusta. Rapporto sulle condizioni di vita e di lavoro dei braccianti stranieri in agricoltura*. www.mediciperidirittiumani.org/pdf/Terraingiusta.pdf

Mingione, E. (1995). Labour market segmentation and informal work in Southern Europe. *European Urban and Regional Studies, 2*, 121–143. https://doi.org/10.1177/096977649500200203.

Mingione, E., & Quassoli, F. (2000). The participation of immigrants in the underground economy in Italy. In R. King, G. Lazaridis, & C. Tsardanidis (Eds.), *Eldorado or fortress? Migration in Southern Europe* (pp. 29–56). London: Palgrave Macmillan.

Moreno, J. (2009). Los contratos en origen de temporada. Mujeres marroquíes en laagricultura onubense. *Revista de Estudios Internacionales Mediterráneos, 7*, 58–78.

Moreno, J. (2012). Movilidad transnacional, trabajo y género: temporeras marroquíes en la agricultura onubense. *Política y Sociedad, 49*(1), 123–140. https://doi.org/10.5209/rev_poso.2012.v49.n1.36525.

Naldini, M. (2003). *The family in the Mediterranean welfare state*. London: Frank Cass.

Natale, F., Kalantaryan, S., Scipioni, M., Alessandrini, A., & Pasa, A. (2019). *Migration in EU rural areas* (EUR 29779 EN). Luxembourg: Publications Office of the European Union. https://doi.org/10.2760/544298.

Nori, M. (2017). Immigrant Shepherds in Southern Europe. *E-paper, Heinrich Böll Stiftung Foundation*. https://www.boell.de/en/agriculture-food-production-and-labour-migration-south ern-europe

Observatorio de las Ocupaciones. (2014). *Datos básicos de movilidad Contratación y movilidad geográfica de los trabajadores en España Datos 2013*. Madrid: Servicio Público de Empleo Estatal.

OECD. (2012). *Indicators of immigrant integration*. Paris: OECD Publishing. https://doi.org/10. 1787/9789264171534-en.

OECD. (2018). *How immigrants contribute to developing countries' economies*. Paris: OECD Publishing.

OPR-Osservatorio Placido Rizzotto. (2018). *Agromafie e Caporalato. Quarto Rapporto*. Roma: Ediesse.

Ortiz-Miranda, D., Moragues, F. A., & Arnalte-Alegre, E. (Eds.). (2013). *Agriculture in Mediterranean Europe: Between old and new paradigms* (Research in Rural Sociology and Development) (Vol. 19). Bingley: Emerald Group Publishing Limited.

Osti, G., & Ventura, F. (Eds.). (2012). *Vivere da stranieri in aree Fragili*. Napoli: Liguori.

Oxfam. (2018). *Human suffering in Italy's agricultural value chain*. Oxfam International & Terra!

Palumbo, L., & Sciurba, A. (2015). Vulnerability to forced labour and trafficking: The case of Romanian women in the agricultural sector in Sicily. *Anti-Trafficking Review, 5*, 89–110. https://doi.org/10.14197/atr.20121556.

Papadopoulos, A. G. (2012). Transnational immigration in rural Greece: Analysing the different mobilities of Albanian immigrants. In C. Hedberg & R. M. do Carmo (Eds.), *Translocal ruralism: Mobility and connectivity in European rural spaces* (pp. 163–183). Dordrecht: Springer.

Papadopoulos, A. G., & Fratsea, L. M. (2016). Appraisal of migrant labour in intensive agricultural systems: The case of Manolada strawberries (Greece). In A. Corrado, C. de Castro, & D. Perrotta (Eds.), *Migrations and agriculture. Mobility and change in the Mediterranean area* (pp. 128–144). London: Routledge.

Papadopoulos, A. G., & Fratsea, L. M., (2017). Temporary migrant workers in Greek agriculture. *E-paper, Heinrich Böll Stiftung Foundation*. https://www.boell.de/sites/default/files/e-paper_ temporary-migrant-workers-in-greek-agriculture.pdf

Perrotta, D. (2014). Vecchi e nuovi mediatori. Storia, geografia ed etnografia del caporalato in agricoltura. *Meridian, 79*, 193–220. https://doi.org/10.1400/221104.

Perrotta, D. (2015). Agricultural day laborers in Southern Italy: Forms of mobility and resistance. *South Atlantic Quarterly, 114*(1), 195–203. https://doi.org/10.1215/00382876-2831632.

Perrotta, D. (2016). Processing tomatoes in the era of the retailing revolution. Mechanization and migrant labour in northern and southern Italy. In A. Corrado, C. de Castro, & D. Perrotta (Eds.), *Migrations and agriculture. Mobility and change in the Mediterranean Area* (pp. 58–75). London: Routledge.

Perrotta, D., & Sacchetto, D. (2013). Les ouvriers agricoles étrangers dans l'Italie méridionale. *Hommes et migrations, 1301*, 57–66. https://doi.org/10.4000/hommesmigrations.1910.

Pigot, M. (2003). *Decent work in agriculture*. http://www.ilo.org/wcmsp5/groups/public/—ed_ dialogue/—actrav/documents/publication/wcms_111457.pdf

Piore, M. J. (1979). *Birds of passage: Migrant labour and industrial societies*. Cambridge: Cambridge University Press.

Piro, V., & Sanò, G. (2016). Entering the "plastic factories". Conflicts and competition in Sicilian greenhouses and packinghouses. In A. Corrado, C. de Castro, & D. Perrotta (Eds.), *Migrations and agriculture. Mobility and change in the Mediterranean Area* (pp. 293–308). London: Routledge.

Prieto, S., & Papadodima, Z. (2010). Reversión (comparativa) del despoblamiento rural a través de las migraciones internacionales. *AGER. Documento de Trabajo n° 28*.

Pugliese, E. (1993). Restructuring of the labour market and the role of Third World migrations in Europe. *Environment and Planning D Society and Space., 11*, 513–522. https://doi.org/10.1068/d110513.

Pugliese, E. (2000). *The Mediterranean model of immigration* (Academicus MMXI/3, pp. 97–106). http://www.academicus.edu.al/nr3/Academicus-MMXI-3-096-107.pdf

Pugliese, E. (2012). *Diritti Violati. Indagine sulle condizioni di vita dei lavoratori immigrati in aree rurali del Sud Italia e sulle violazioni dei loro diritti umani e sociali*. Dedalus Cooperativa Sociale.

Pumares, P., & Jolivet, D. (2014). Origin matters: working conditions of Moroccans and Romanians in the greenhouses of Almeria. In J. Gertel & S. R. Sippel (Eds.), *Seasonal workers in Mediterranean agriculture. The social cost of eating fresh* (pp. 130–140). New York: Routledge.

Ragkos, A., Koutsou, S., & Manousidis, T. (2016). In search of strategies to face the economic crisis: Evidence from Greek farms. *South European Society and Politics, 21*, 319–337. https://doi.org/10.1080/13608746.2016.1164916.

Reigada, A. (2016). Family farms, migrant labourers and regional imbalance in global agri-food systems. On the social (un)sustainability of intensive strawberry production in Huelva (Spain). In A. Corrado, C. de Castro, & D. Perrotta (Eds.), *Migrations and agriculture. Mobility and change in the Mediterranean area* (pp. 95–110). London: Routledge.

Salazar Parreñas, R. (2001). *Servants of globalization*. Stanford: Stanford University Press.

Sampedro, R. (2013). Spatial distribution of foreign labor immigrants in rural areas: Exploring the potential of towns and villages to retain them in the long run. In AA.VV. (Ed.), *Proceedings of the XXV conference of the European Society for Rural Sociology*. Pisa: Laboratorio di studi rurali Sismondi.

Sampedro, R., & Camarero, L. (2015). *International immigrants in rural areas: The effect of the crisis in settlement patterns and family strategies*. Proceedings of the XXVI Congress of the European Society for Rural Sociology, Aberdeen (UK).

Sanò, G. (2018). *Fabbriche di plastica. Il lavoro nell'agricoltura industriale*. Verona: Ombre Corte.

Schmoll, C., Thiollet, H., & Wihtol de Wenden, C. (Eds.). (2015). *Migrations en Méditerranée/migration in the Mediterranean*. Paris: CNRS editions.

Schrover, M., Van der Leun, J., & Quispel, C. (2007). Niches, labour market segregation, ethnicity and gender. *Journal of Ethnic and Migration Studies, 33*(4), 529–540. https://doi.org/10.1080/13691830701265404.

SOFA. (2018). *State of food and agriculture 2018 on migration, agriculture and rural development*. Rome: FAO, Food and Agriculture Organization. http://www.fao.org/3/I9549EN/i9549en.pdf.

Solé, A. (2010). Características sociodemográficas, pautas de distribución territorial y proceso migratorio de la población de nacionalidad extranjera en el Alt Pirineu i Aran: contribuciones a la transformación de un espacio de montaña. *AGER. Documento de trabajo*, 290.

Triandafyllidou, A. (2013). *Circular migration between Europe and its neighbourhood: Choice or necessity?* Dordrecht: Oxford University Press.

UN DESA. (2017). *Trends in international migrant stock: The 2017* revision (United Nations database, POP/DB/MIG/Stock/Rev.2017). New York: United Nations, Department of Economic and Social Affairs.

Vaiou, D., & Hadjimichalis, K. (1997). *With the sewing machine in the kitchen and the poles in the fields: Cities, regions and informal labour*. Athens: Exandas.

Viruela, R., & Torres, P. F. (2015). Flujos migratorios, crisis y estrategias de movilidad. Los inmigrantes ecua- torianos y rumanos en España. In F. Torres Pérez & E. Gadea (Eds.), *Crisis, Inmigración, Sociedad* (pp. 37–72). Madrid: Editorial Talasa.

Zuccotti, C. V., Geddes, A. P., Bacchi, A., Nori, M., & Stojanov, R. (2018). *Rural migration in Tunisia. Drivers and patterns of rural youth migration and its impact on food security and rural livelihoods in Tunisia.* Rome: Food and Agriculture Organization of the United Nations.

Chapter 4
Rural Destination Areas: Impacts and Practices

In this chapter we provide a framework to assess and analyse ongoing rural migration dynamics from the perspective of areas of destination, with a view to answer to the following questions: What are the impacts on the local economy and society? Which are the practices, programs and policies that underpin the presence and integration of migration? What is recent experience revealing on these matters?

In particular, we focus on the more marginal, isolated, remote areas of the EUMed where the contributions of immigrants are critical for the sustainability and reproduction of local societies. In these areas, immigrant communities represent a strategic asset with a vision to contrast processes of population decline and overall socio-economic desertification. The chapter progresses through several cases and experiences related to processes and practices of inclusion and integration of immigrants in diverse rural settings in Italy.

4.1 Introduction

Chapter 3 discusses the important role of immigrant workers in Mediterranean agriculture. In this chapter we will analyse the consequences of immigration in different areas of destination, while in the following chapter those of emigration on the communities of origin will be assessed.

Existing literature mostly focuses on the migrant workforce employed in EUMed intensive agricultural systems, oftentimes addressing its exploitative nature and relationships (King et al. 2000; Ortiz-Miranda et al. 2013; Gertel and Sippel 2014; Corrado et al. 2016, 2018; Corrado 2017; Papadopoulos and Fratsea 2017). This literature plays an important role in raising concern over the (often degraded and vulnerable) living and working conditions of rural migrants, and the related economic, social and political implications of such arrangements.

As the focus is mostly on agricultural high-potential areas and intensive systems, the relevance of the migratory phenomenon in agro-ecological marginal settings has

© The Author(s) 2020
M. Nori, D. Farinella, *Migration, Agriculture and Rural Development*, IMISCOE
Research Series, https://doi.org/10.1007/978-3-030-42863-1_4

been often overlooked by academic literature. This is due to several reasons, including the marginality of these territories and their limited relevance in policy debates. Nevertheless, it is particularly in these marginal rural settings that the presence and contributions of migrant communities are critical in maintaining these territories and ecosystems alive and productive. Here we mostly focus on these latter areas, as a way to redress existing literature, and also because immigrant communities represent there a strategic asset to contrast rural population decline and shrinking agricultural practices.

This chapter provides a framework to assess and analyse ongoing rural migration dynamics in either settings, with a view to answer to the following questions: What are the impacts on the local economy and society? Which are the practices, programs and policies that underpin the presence and integration of migration? What is recent experience revealing on these matters?

These questions will be addressed for the different agricultural and rural development patterns pertaining to systems in EU Mediterranean countries, with a specific focus on Italy, as emblematic case for these dynamics. Related experiences and initiatives aimed at creating local models of integration in either intensive and extensive settings will be then assessed, towards more sustainable agricultural productions systems and rural development patterns.

4.2 Implications in Rural Areas of Destination

As it has been discussed in the previous chapters, the decline and ageing of population that have characterised rural settings in recent decades have resulted in problems of workforce availability and generational renewal. These problems have been threatening the sustainability of agriculture, food systems and rural communities alike in parts of Europe (Nori 2017a; Farinella et al. 2017; FAO 2018).

As we have analysed, immigrant populations have often come to replace and complement the declining local one, with evidence attesting that in most cases, the immigrant labour force does not compete with native workers, but it rather fills the gaps in agricultural labour markets (Kasimis 2010; Nori 2015; FAO 2018; Robinson et al. 2017).

As discussed in Chap. 3 the geographical features of Mediterranean countries together with the socio-economic and territorial polarisation that has characterised rural development in recent decades have contributed reconfiguring the agrarian world in two basic domains:

1. Labour-intensive farming, livestock breeding and horticultural value chains that characterise agricultural systems in high potential areas such as valley bottoms, plains and coastal areas.
2. Low-input systems and agro-pastoral practices in marginal rural settings that offer limited capacities for agricultural intensification—mountainous areas, remote villages, islands.

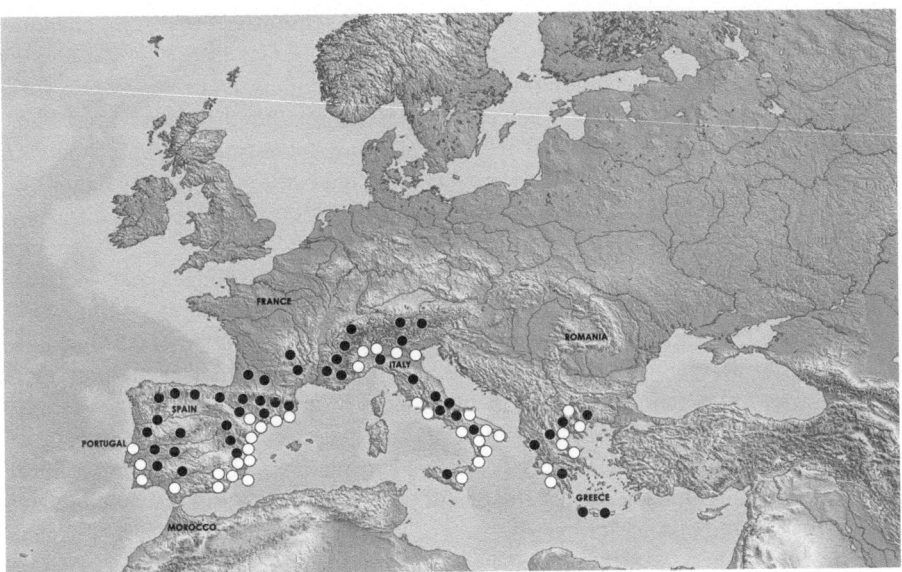

Fig. 4.1 Main areas of intensive agriculture (white) and extensive agro-pastoralism (black) in EUMed countries. (Source: our elaboration)

Figure 4.1 indicates main areas of intensive and extensive farming systems in the EU Mediterranean region.

The presence and contribution of immigrant communities is widespread in both settings and in related farming systems; the patterns, dynamics and challenges are though quite diverse. These have given rise, through time, to a territorial reconfiguration along forms of ethnic specialisation, with distinct communities occupying specific ecological and productive rural niches (Schrover et al. 2007; Bell and Osti 2010; Ambrosini 2013).

Migrants in high-potential areas satisfy the high demand for temporary, cheap and precarious labour which is demanded by such seasonal and intensive systems, which require workers to move from one region to another according to production needs. These contributions and impacts have been thoroughly assessed in the previous chapter.

In marginal rural areas the human presence goes beyond the mere economic dimension, as it bears relevant implications in the social and environmental dimensions as well. As it will be assessed, land abandonment and the reduction of ecosystem management associated to local farming, grazing and forestry implies substantial natural hazards for society. Furthermore, the active presence of people in marginal settings has a wider, 'multifunctional' role in maintaining local territories and reproducing local societies. These are the reasons why the social desertification most marginal rural communities have undergone in recent decades represent a serious concern for policy makers and citizens alike. The growing presence of foreign, immigrant communities represents in these settings an interesting phenomenon, which might help to redress these dynamics.

Research suggests that immigrant communities residing and operating in marginal settings not only represents a main supplier of agricultural-related wage labour, but oftentimes play a plurality of roles, alternating between agriculture, tourism, construction and other service provisions which often carry relevant implications on the local social and cultural fibre (Mas Palacios and Morén-Alegret 2012; Kasimis and Papadopoulos 2013; Nori and Luisi 2019).

Their contributions are often vital to rural enterprises, villages and societies, which in the last decades have suffered from problems associated to lacking workforce, ageing population and generational renewal. The contribution of immigrants is also critical in providing the ecological and social services that support local societies, as much as it is in mere demographic terms.[1] Through higher fertility rates immigrant communities play a relevant role in supporting local demography by buffering population decline. This has helped to maintain the provision of basic services, such as primary schools and health posts, for remote and poorly populated areaswhich have been the primary targets of shrinking public budgets due to their lowering population density and limited political influence (Kasimis et al. 2010; Osti and Ventura 2012; Gallo and Rioja 2016).

The demographic structure, average age and fertility rates of immigrant communities compared to local ones suggests that their relevance for the social and economic development of these areas will be increasing through time. As for Italy, immigrants represent about 10% of the adult population in most inner areas of central Italy (refer to Table 4.1), where the proportion of children in local schools in normally much higher, thus to indicate a shift in the local population composition (Barca et al. 2014; SNAI 2015; Nori and Luisi 2019). Similar dynamics are reported for island and mountainous areas of Greece and Spain (Kasimis 2010; Collantes et al. 2014).

Immigrant women in these settings often play less visible but equally critical roles by providing domestic work and care-giving services. Oftentimes the availability of foreign assistants enables local elders to remain inhabiting rural villages, while allowing their relatives to engage in the labour market. In these terms immigrants complement local labour, in the fields and at home, facilitating the adoption of new employment strategies of autochthonous families, with overall positive effects on the local economy (De Lima et al. 2005; Kasimis et al. 2010; Kasimis and Papadopoulos 2013; Osti and Ventura 2012; Mas Palacios and Morén-Alegret 2012; Nori and López-i-Gelats 2017; Ragkos et al. 2018).

---

[1]Refer to data from Kasimis 2010; Collantes et al. 2014; Barca et al. 2014; SNAI 2015 respectively for island and mountainous areas of Greece, Spain and Italy.

Table 4.1 Population growth rate in Italy: annual average in inner areas municipalities (years 2003–2013 × 1000 inhabitants)

Regions of Italy		Total population annual average growth rate	Italian population annual average growth rate	Foreigners population annual average growth rate
Islands	Sicily	−0.15	−2.35	+2.18
	Sardinia	0.76	−1.09	+1.82
Alpine inner areas	Piedmont	0.63	−4.16	+4.63
	Aosta Valley	2.77	−1.76	+4.39
	Lombardy	3.44	−1.48	+4.59
	Trentino Alto Adige	6.48	2.79	+3.49
	Veneto	4.34	−0.94	+4.95
	Friuli Venezia Giulia	−2.03	−5.61	+3.49
Apennines inner areas	Liguria	−0.48	−4.76	+4.19
	Emilia Romagna	3.77	−1.84	+5.29
	Tuscany	1.55	−3.92	+5.32
	Umbria	3.95	−1.72	+5.56
	Marche	−0.80	−5.62	+4.59
	Lazio	13.31	6.25	+6.97
	Abruzzo	−0.51	−4.16	+3.61
	Molise	−3.55	−6.01	+2.44
	Campania	−1.41	−4.06	+2.63
	Apulia	−0.53	−2.33	+1.77
	Basilicata	−4.72	−6.82	+2.09
	Calabria	−3.03	−5.93	+2.87

Source: Own elaboration based on Istat data (Nori and Luisi 2019)

Box: Romanian Women in Rural Sardinia

Romanians are the main foreign residents in Sardinia; of a totality of 14.216 in 2018, Romanian women (9.626) are double than men (4.590). Most Romanian women work in rural and inner mountainous areas such as in Barbagia, where they assist the elderly as caregivers. There are frequent cases of inter-marriages between locals and Romanian women who eventually contribute to the farm management and economy. However, there are no real policies to support the inclusion of these women in local societies. The general attitude seems rather that of an opportunistic exploitation of cheap labor to compensate for the shortcomings of a welfare state and the absence of care policies.

(continued)

A similar phenomenon was reported in the early 2000s when northern Greek mountainous communities received important inflows of Albanian immigrants, following the change of policy regimes (Kasimis 2008, 2010).

In comprehensive, broader terms, immigrants' contributions in demographic, economic and social terms play an important role in covering the gaps created by an ageing society and the associated decline in welfare services (*for more references* see AA.VV. 2018). All in all their presence is therefore critical in maintaining and reproducing local communities and the socio-cultural identity of territories (Kasimis et al. 2010; Osti and Ventura 2012; Barca et al. 2014; SNAI 2015; Desjardins et al. 2016; Gallo and Rioja 2016; Nori 2017b).

In Chap. 6 immigrants' role in ensuring basic ecological services will also be assessed; the specific agro-pastoral domain will be discussed in deeper detail, as a case study to disentangle and assess these contributions to the sustainable development of marginal territories.

The following two sections assess several experiences related to processes and practices of inclusion and integration of immigrants in diverse rural settings in Italy. We consider the Italian case as an example of the initiatives and debates underway in Mediterranean Europe on these issues. The choice to focus on this country is linked to a better knowledge of the reference literature and experiences, and enables the author to provide a comnprehensiv overview of the range of initiatives undertaken at national level.

4.3   Experiences in Integrating Migrants in Intensive Agricultural Areas

Several experiences and practices have been set up and evolved in recent years in most EUMed settings with a view to contrast the exploitative local conditions of immigrant workers. Civil society has been active in proposing and implementing bottom-up practices of integration and cooperative agriculture where local farmers, activists, precarious workers (both local and migrants) and refugees jointly engage to contrast the agricultural squeeze of producers and the exploitation of workers. We will assess a number of these pertinent to the Italian context.

Through the promotion of alternative agriculture practices, based on fair relations and short supply chains, these bottom-up experiences aim to promote better living and working conditions for agricultural workers, and to support their inclusion in the local society, addressing "simultaneously the crisis of social reproduction of both small-scale farmers and of migrant farm workers" (Iocco et al. 2017). These experiences are limited, but significant because they try to propose an alternative model, based on peasant agriculture and on the alternative food networks and short supply chains. The idea is to promote an agriculture reconnected to the local environment, which is able to give value to the reciprocity of relationships, more oriented towards a "moral economy".

The focus of these experiences is to overcome a model in which agriculture is only conceived as a system to produce commodities, to be sold at low cost on the global market. Instead agriculture is here conceived and practiced as a multifunctional activity aiming to ensure a fair standard of living for local people, to improve social and fair relations between all actors in the agri-food chain based on local knowledge, respect of the local ecosystems and mutual relationships, while guaranteeing the quality of healthy food for consumers. The political framework most civil society initiatives breed and evolve from is the one articulated at the global scale through the *Via Campesina* network (https://viacampesina.org/en/).

SOS Rosarno operates in an area of Calabria that is characterized by an intensive exploitation of immigrant workers engaged in the citrus fruit harvest that determined, in 2010, a first immigrant strike to obtain rights to a fair salary. Despite the symbolic significance of this uprising, the situation of migrants in Rosarno is critical, with migrants forced to live in slums and exploited in the local countryside, with strong episodes of racism by the local residents. In this difficult context, in the 2011, an organic farmers' cooperative, *I frutti del sole*, started to employ four African workers in the harvest of citrus with regular contracts and fair retribution, organising the commercialisation of the citrus fruits in some critical consumers' networks (whom are able to support the ethical aims of the project), in particular through solidarity purchase groups based in Rome, Bologna and other main Italian towns. This first project was extended in 2012 with the foundation of SOS Rosarno, an association for social development created by local farmers, activists and African workers to promote the original idea of a transparent and fair citrus fruits chain (also refer to www.sosrosarno.org and Oliveri 2015; Mostaccio 2013; Iocco and Siegmann 2017; Semprebon et al. 2017; Iocco et al. 2018, 2019).

SOS Rosarno was launched as a solidarity economy project, with the aim to promote the collaboration between local farmers and African workers around an alternative citrus fruit chain, based on the right to the fair remuneration for all participants in the supply chain (farmers, workers, final consumers). Although it was a successful project, this experience had a weak point: the precariousness of the work. Though fair salary and fair working conditions have been guaranteed, labour availability remained seasonal, linked only to the harvest of citrus fruits (and to the unemployment benefits in the rest of the years). As Iocco and Siegmann (2017) noted, this did not allow the immigrants to send remittances home and fulfil a central aspect of their migratory project.

For this reason, recently the project involved with the social cooperative *Mani e Terra* (Hands and land), in which the migrant workers themselves are members. The idea is increasing the empowerment of the local and immigrant workers, involving them in the farm's management, processing also other vegetables and other agricultural products (to give a guarantee of a continuity of salary to the working member all of the year) and experimenting some practices of collective farming, through the rent and the cultivation of some hectares of abandoned land.

The project called *Funky Tomato* (FT) started in 2015 and focused on the ethical harvesting of tomatoes in the area of Venosa (Basilicata Region) and their artisanal processing of canned tomatoes, that are distributed in alternative and critical food networks as solidarity purchase. It is inspired by the informal group *Fuori dal ghetto* (Out of the ghetto) which aims to support the immigrant workers in the tomato harvest to obtain a fair retribution and dignified living conditions (out of the "ghetto" of Venosa) (Iocco et al. 2018, 2019). Small farms and social cooperatives participating in the project accept Funky Tomato ethics and principles, and engage with FT agricultural company in order to produce process and commercialise tomatoes along the FT lines. The project has continuously grown since its inception, increasing the tomato cans production and the amount of workers employed. From its origin in Basilicata, it eventually extended as well to the regions of Sicily and Campania (Iocco et al. 2017).

Similar to Funky Tomato is the project *SfruttaZero* in Apulia (Italy). This is a cooperative and mutualistic project that aims for the cooperation of migrants, farmers, young people precarious in agricultural activities to produce local products (in particular tomato preserves), to be sold in solidarity economies. This experience was born after the strikes of the agricultural workers of Nardo and aims to enhance diversity strengthen social relations and fight against exploitation (Perrotta and Sacchetto 2015).

Other similar experiences are made by the social Cooperative GOEL in the Locride area (Calabria), a region with a high rate of organized crime. The cooperative promotes a series of activities based on the ethical principles of solidarity, fair pay, the fight against crime and the enhancement of local knowledge and cultures. In this context, it has launched a project for the collection and marketing of oranges in circuits of ethical economy, which guarantees a fair remuneration to the workers employed in the collective (migrants and locals) and to farms.

Another interesting micro experience of inclusive and ethical production has been elaborated by the *Barikamà cooperative* which was created following the struggles of Rosarno thanks to a group of African workers. After several experiences of exploitation in the Calabrian countryside for the harvest of oranges and in those of Puglia for the harvesting of tomatoes, in 2012, young Africans founded the association that produces yoghurt. This project started in the premises of a former social center in Rome, which also started a microcredit project to finance activities that involve African workers, originating from Mali, Senegal, Benin, Gambia, Guinea, of which four are involved in the Rosarno uprising (Ascione 2018).

Contadinazioni is a movement localized in the area of Mazara (Sicily), where African workers are exploited in the olive harvest in Autumn season and live in rural shanty towns without any basic services (water, electricity, toilets, etc.) (Iocco et al. 2018, 2019). In the 2013, a young migrant worker died in his shack due to an explosion. A collective of local activists, militants and researchers based between Palermo and the small village of Campobello di Mazara, mobilized to help the migrants living in the slums by improving access of water and by promoting mutualistic relationships to self-produce goods and services useful to improve the quality of life of immigrants in this territory. The idea of this movement is to promote self-production, mutualism, cooperation, and peasant practices as social cohesive experiences targeting young people, and local and immigrant communities.

Members of the Contadinazioni started to work in seasonal olive harvesting to improve their know-how in the agricultural sector (many of them have not experience in agriculture and grew-up in an urban context) and to create linkages with African workers and olive producers. After this phase, Contadinazioni started an autonomous experience of production of organic table olives, in collaboration with SOS Rosarno, with the idea to push-back against the exploitation of immigrant people. Many Senegalese workers are part of this movement.

The project also succeeded to establish an agricultural cooperative *Terra Matta,* by uniting a few migrant workers, young local precarious workers and activists, with the aim to promote sustainable agriculture and collective production of sun-dried tomatoes. This collective is very fluid and heterogeneous, characterized by a strong political component. The participatory and deliberative mechanisms of decision making are particular to this project (Iocco et al. 2018). Though, in some cases they slowed down the decision-making process and created conflict.

The *Maramao* project (Donatiello and Moiso 2019), financed in part by funds from the Protection System for Asylum Seekers and Refugees (SPRAR), managed by the Ministry of the Interior, is interesting because is located in the North of Italy (Piedmont Region), in an area with intensive agricultural development, in Canelli town. Canelli is specialized in the cultivation of the vineyards from which the Moscato Bianco DOCG vine is produced. This area is also recognized by UNESCO as a World Heritage Site for its extensive vineyards that characterize the rural landscape. The paradox is that the Moscato wine, a typical and local product of the territory, is produced through the exploitation of migrants, hired through the mechanism of landless cooperatives. Landless cooperatives often provide for the intermediation of other migrants, thus increasing local migrant populations. For example, the Macedonian community now settled in the area, accounts for 8% of Canelli's population (Donatiello and Moiso 2017, 2018).

The Maramao project aims to facilitate immigrant entrepreneurship by supporting agricultural production, in the name of enhancing the quality and the link with the territory. The involvement of migrants takes place through the SPRAR that involves refugees and asylum seekers (Zetter 2017). At its base there is an agricultural cooperative that cultivates abandoned farmland and is free or cheap to some residents. The cooperative also carries out training and job placement activities for young migrants from neighboring SPRAR. Currently, its staff is made up of five people, including three refugees carrying out agricultural activities and one Italian person presenting a percentage of disability. There are also five other trainees, including four asylum seekers and holders of international protection (Donatiello and Moiso 2019).

The *Sicily Integra* project was established at the end of 2015 by an NGO, with the aim of promoting equitable and sustainable development and the active inclusion of local young people and migrants. It has foreseen the activation of local training projects on the sustainability of biological and regenerative agri-food systems and on sustainable agriculture (fair-farms, ecovillages, agro ecological movements, farming techniques, etc.), reaching 93 people, of which 23 young unemployed Sicilians and 70 migrants, asylum seekers and refugees entering the Italian SPRAR system (Dara Guccione et al. 2018).

In other EUMed countries we find other similar projects. In Spain civil society actions include the experience of the *Sindacato de Obreros del Campo* (SOC) which is active in Andalusia in integrating local claims on land and on labour rights with support to the integration of foreign workers (Caruso 2016). The *Nuevos Senderos* program is engaged in supporting the integration of immigrant households in rural communities suffering from intense depopulation (www.cepaim.org; http://nuevossenderos.es/). In Greece and Turkey as well civil society organizations are actively engaging in supporting the integration of Syrian refugee agricultural workers by enhancing their access to farm land and improving the recognition of their rights (such as the Development Workshop initiative: http://www.ka.org.tr/).

Many of these experiences are micro and bottom-up practices of social innovation and inclusion based on the cooperation between immigrant workers and local people to promote new forms of peasant agriculture, with a mutualistic approach, linked to economies of solidarity and reciprocity, and models of critical consumption as the short food supply chains and the alternative food networks.

Characteristic features common to most schemes include:

- the organization in a cooperative or associative form;
- the importance of the reuse of land previously abandoned and often worked collectively;
- the valorization of local knowledge and the training of newcomers on local techniques and practices;
- the building of networks of sharing, exchange and co-production between migrants and locals in a relational economy;
- the attention to fair prices in the supply chain for workers and farmers;
- the reliance on critical/ethical consumption demands that recognize and remunerate the material, ecological and cultural value of productions;
- the central role of militants and local activists in supporting initiatives;
- the choice to encourage a peasant agriculture that invests in the cultural and social values of food and agriculture, with a view to enhance organic production and workers' rights.

Most initiatives hold the merit of trying to build horizontal and bottom-up cooperation practices, with a view to enhance the subjectivities of the population present in the territories involved, both migrant and native. One of the most interesting aspects is the strengthening of the autonomy through community practices and collective action, starting by overcoming the role of subordinated worker and recovering the ethical and moral dimension of agriculture. Another relevant aspect is the ambition to build economies of reciprocity and mutual-aid, not simply focused on economic value but embedded in social relationships. However, these processes (based on deliberative democracy mechanisms) often proceed through tiring and conflicting paths and the related principles that underlie these experiences are time-demanding, controversial and stressful; it is often difficult to balance the interests and motivations of the different participants.

4.4 Experiences in Integrating Migrants in Marginal Rural Settings

Several programs and experiences have been implemented with the view to integrate immigrant communities in marginal rural settings of Spain, Italy and Greece. Most initiatives have though originated from the need to allocate the intense flows of refugees and asylum seekers generated by the political and economic crises that ravaged the Mediterranean, rather than by a genuine concern for inclusive and sustainable agricultural systems and rural communities. We will present here cases from the Italian context, through governments programs as well as civil society actions in southern and northern parts of the country.

An important experience along these lines is the *National Strategy for Inner Areas* (SNAI), a long-term strategy financed by EU and through national funds with the objective to counter the socio-economic marginalization of inner areas of the country, which represent about three fifths of the Italian territory. These areas are characterized by a lack of basic services and depopulation dynamics, demographic malaise. The challenge in these areas is to trigger local dynamics of social and entrepreneurial vitality, to improve the quality of life and the related attractiveness of these areas, with the aim of reversing ongoing trends of rural desertion (Nori and Luisi 2019).

The strategy is based on the idea that economic development and social cohesion are objectives to be achieved jointly, working on intertwined aspects. For example through the improvement of citizenship services (school, health, mobility), enhancement of environmental diversity and landscape, inclusion of projects aimed at countering the hydro-geological and landscape instability, the strengthening of the territorial capital, and construction of new connections and circularity between urban and rural areas (refer to http://community-pon.dps.gov.it/areeinterne/).

This strategy proposes support for new forms of rurality and the so-called new rural populations. Amongst the agents of change critical to this challenge—which includes young and new farmers, retired people, temporary residents—immigrant communities who work and live in rural areas, as well as those who are hosted in refugees and asylum seekers reception facilities, represent an important potential asset. These actors carry skills and capacities that would be critical to enhance local, endogenous development.

Box: Migrants' Work Contributing to Managing Natural Parks (Nori and Luisi 2019)

In the twentieth century Italian silviculture took shape in the forested areas of Casentino, in Tuscany. These practices eventually influenced forest management in the whole Mediterranean and beyond. These areas are today protected and enhanced through the establishment of a National Park. In the Municipalities of the Casentino Park, the presence of immigrants is the highest among Italian national parks (12.3%). A significant part of these neo-citizens come from the Bakau region of Romania, rural areas that have many similarities with the local ones. This is a main reason the experience and the technical skills of Romanian foresters are recognized and appreciated, and represent a resource for the local territory. The immigrant population plays therefore a fundamental role in the conservation and evolution of the forest sector—which is at the same time a landscape heritage, a touristic attraction and an element of the local traditional identity, and a key resource for the park as well as for the future of this territory.

Since the beginning of the 2000s, the Italian government set up a national system to receive refugees based on a private-public partnership, specifically between local authorities and civil society actors. This system is named *Sistema di protezione per richiedenti asilo e rifugiati* (SPRAR—Protection systems for asylum seekers and refugees), and receives funding from the Ministry of the Interior. A main pillar of this system is SPRAR's hosting centers spread out throughout the territory, particularly in rural and inner areas. The vision is to avoid the territorial concentration and the "ghettoization" of migrants, while exploiting the availability of housing in these areas.

Through improved reception of migrants and refugees, the SPRAR project hopes to generate social capital, relational goods and external economies useful for the growth of the territory as a whole. To this end SPRAR aims to:

– develop widespread hospitality paths for migrants, in which migrants can become main actors in the revitalization of depopulated villages (through projects for the recovery and management of abandoned houses where migrants would be hosted for example);
– favour training in liaison with the local population to enhance exchanges, sharing and ultimately integrate and collaborate on the provision of local public goods and/or services to enhance rural welfare.

The general approach is based on the construction of a process of community empowerment and community care, in which the local population and refugees experimented with forms of active citizenship, participation, continuous learning, with the aim of producing developing that relational and cultural fabric in able to support endogenous local development based on social cohesion and territorial innovation.

The establishment of *Centri di Accoglienza Straordinaria* (CAS—Extraordinary Reception Centers) has been informed by emergency, a humanitarian logic, rather than by aspects of active inclusion of refugees and their potential contribution to the revitalization of local societies.

Systems of refugees and asylum seekers reception have had variable and diverse outcomes and impacts. In some cases, these have generated virtuous processes of social innovation and territorial revitalization, while in others they have resulted in precarious livelihoods, patterns of exploitation and social tensions (Corrado et al. 2018 also refer to Chap. 3). In general civil society has been critical for enhancing the capacity of EUMed territories to integrate and include newcomers in recent decades. In Italy this is specifically the case for mountainous communities, where estimates indicate the number of foreigners to be around 350,000; this includes important portions of asylum seekers and refugees. About 30% of total refugee flows have been relocated '*either by choice, by force or by necessity*', in mountainous communities (Membretti et al. 2017). Successful strategies to enhance integration have targeted both immigrant and local populations, often working and investing at the interface amongst them, and creating opportunities for synergies and cooperation.

A well-known program of reception and insertion of immigrant communities is that of **Riace**, a small village in Calabria, southern Italy. When in the late 1990s the village witnessed the arrival of hundreds of Kurdish refugees that approached Calabria coasts through boats, the mayor Domenico Lucano decided that was a relevant opportunity to welcome and host them, while also envisaging a different future for a community that had been undergoing decades of population decline and economic depression (Sarlo and Martinelli 2016; Carrosio 2019).

The whole administration and management of Riace was devoted to integrating these new citizens, in a proactive way, through recovering the ruinous local building heritage, opening schools, financing micro-activities, opening bars, artisan workshops, bakeries, and shops, reclaiming lands and agricultural production and setting up new services to this aim. Apart from providing a livelihood to incoming populations, these opportunities generated local employment and supported the local economy, which eventually benefitted to the whole population. Cultural mediators were employed and involved in the process, and eventually innovative touristic tours attracted European visitors to witness the experience of the 'capital of hospitality'.

The town tripled its inhabitants, eventually turning from a ghost town, to a vibrant social, economic and cultural centre, providing hospitality to over 6000 asylum seekers from more than 20 countries. Riace became a model, showing that fair integration is beneficial to everybody, and indeed indicating a potential scheme for inspiration and replication. In 2010 the film maker Wim Wenders decided to film this experience through the movie '*Il volo*'. Along these principles, similar SPRAR projects eventually developed locally in Badolato, Caminiti, Caulonia and Stignano (Carrosio 2019).

Other interesting example is that of Pettinengo, Piemonte, which is reported, amongst others, in Perlik and Membretti (2018: 258). Set in an area which had undergone a deep socioeconomic and identity crisis, with persistent negative natural demographic balance, *Pacefuturo* launched in 2008 the project "Sent-ieri, oggi e domani" (Pathways—yesterday, today and tomorrow). With the view to support the local integration of asylum seekers and refugees. The NGO in collaboration with the municipal administration and with the active involvement of the local community, the project brought back to life more than 10 km of old paths that connected the farms and the larger neighbourhoods of the village. These paths were used by peasant workers to reach the sites of now-abandoned factories.

The aim of the project was to appreciate the natural and cultural landscapes crossed by these paths, countering the abandonment of the area. The strategy combined the local need for restoring the cultural heritage with the need expressed by asylum seekers for concrete opportunities of inclusion in the community and in its territory. The migrants were enrolled as members of the association, and contributed as volunteers for the maintenance of the landscape. While working in the field, the migrants also received technical training, often provided by local people.

By combining cultural growth, the development of tourism, and social solidarity, the project has promoted the transformation of an area afflicted by negative social and economic trends. From its beginnings, the municipal administration of Pettinengo has actively supported Pacefuturo, while requiring that every service offered to newcomers must also be offered to the entire population. Thus, the original residents also benefitted from the services offered to refugees. The project has eventually become a breeding ground for further projects aimed at including refugee populations.

Other interesting experiences are reported from mountainous Alpine settings that are common to several southern European countries, which have also recently witnessed an important shift in local human and cultural landscapes. Examples can be drawn from the work of Perlik and Membretti on *Alpine Refugees* (2018), and by the projects PADIMA (www.padima.org), or PlurAlps (https://www.alpine-space.eu/projects/pluralps/en/home).

In 2015 a wide debate sparked in European countries about refugees' accommodation in rural areas. In the new rural development plan of the EU this strategy is considered an opportunity in face of the high population density of urban areas and the rural exodus which increases the availability of housing in these areas. In this section we have explored some Italian practices as examples of the issues at stake, but a wide literature involving other EU countries is available (see ENRD 2016; Papageorgiou et al. 2016; Scholten et al. 2017; Galera et al. 2018; Weidinger 2018). The refugees' accommodation in rural areas can be an opportunity to revitalize the economic and social fabric, contrasting the decrease of services in rural areas (stimulated by the "newcomers"). However, research on these experiences provides controversial results. Some micro-experiences of widespread hospitality have worked, while others have been found to be conflictual and problematic, with a rejection by the local population and effects of alienation on the migrants'side (who have felt more isolated and victims of a double peripheralization).

4.5   Conclusions

The socio-economic and territorial polarisation that has characterised rural development in recent decades have contributed reconfiguring the agrarian world into a) areas of intensive agriculture, characterised by a high demand for cheap, temporary and precarious workforce, and b) extensive agricultural practices in more marginal rural settings, characterised by local population decline.

Though with different dynamics foreign communities who immigrated in rural areas have come to fill the gaps left by the local population in these rural settings. Evidence shows that their contributions have enabled many farms, rural villages, and agriculture enterprises to remain alive and productive throughout the recent financial crisis, thus representing a critical asset in enhancing the resilience of the European rural world.

Nevertheless immigrants' presence, conditions and integration in rural areas represent a matter of concern at different levels. As described practices and experiences exist which aim to recognize and improve the rights, working and living conditions of rural immigrants and that support their integration in the local society, with positive effects for the local and the immigrant populations alike. These are all parts of the same challenge towards more sustainable models of agriculture and rural development.

Most positive experiences of migrant inclusion are linked to the idea of a different model of agriculture and alternative food networks. The rural immigrant question must become part of the European debate between a multifunctional agriculture in opposition to a modern one, in terms of efficiency, performance as well as sustainability, with relevant implications for the principles underpinning policy frameworks, including the CAP (Van der Ploeg 2008; Marsden and Franklin 2013; Corrado et al. 2018).

We need though to get out of a hetero-direct policy that forces immigrants to live and stay in often disadvantaged rural areas, producing a ghettoization effect. This is one of the risks these projects aiming to accomodate asylum seekers and refugees in rural settings without directly involving all local stakeholders. On the one hand, immigrants are forced to go to areas where they are hosted in structures separated from local communities, creating a double spatial segregation. On the other hand, the rural space is once again imagined as "empty", "to be filled", ignoring the presence of local communities and ongoing dynamics.

In conclusion, this debate should consider and include the improvement of rural migrants' living and working conditions as the necessary step to shift from the dimension of "workers" or "refugees" to that of "citizens".

References

AA.VV. (2018). Welfare Oggi, 5. http://www.welfareoggi.it/

Ambrosini, M. (2013). *Irregular migration and invisible welfare.* London: Palgrave Macmillan UK.

Ascione, S. (2018). La cooperativa Barikamà: dallo sfruttamento delle campagne all'autogestione del lavoro. In RRN (Rete Rurale Nazionale). *Terreni d'integrazione, 3*(31), 44–47.

Barca, F., Casavola, P., & Lucatelli, S. (Eds.). (2014). *A strategy for inner areas in Italy: Definition, objectives, tools and governance* (Materiali Uval Series) (Vol. 31). Rome: Uval, Agenzia per la coesione territoriale.

Bell, M. M., & Osti, G. (2010). Mobilities and ruralities: An introduction. *Sociologia Ruralis.* Special Issue on Mobilities and Ruralities, *50*(3), 199–204, https://doi.org/10.1111/j.1467-9523.2010.00518.x.

Carrosio, G. (2019). Migranti, mercati nidificati e sostenibilità in territori fragili: i casi di Riace e Camini (Calabria). *Mondi Migranti.* Forthcoming.

Caruso, F. (2016). Unionism of migrant farm workers. The Sindicato Obreros del Campo (SOC) in Andalusia, Spain. In A. Corrado, C. De Castro, & D. Perrotta (Eds.), *Mobility and change in the Mediterranean area.* London/New York: Routledge.

Collantes, F., Pinilla, V., Sàez, L. A., & Silvestre, J. (2014). Reducing depopulation in rural Spain. *Population, Space and Place, 20*, 606–621. https://doi.org/10.1002/psp.1797.

Corrado, A. (2017). *Migrant crop pickers in Italy and Spain* (E-paper), Berlin: Heinrich Böll Stiftung Foundation. https://www.boell.de/sites/default/files/e-paper_migrant-crop-pickers-in-italy-and-spain.pdf

Corrado, A., De Castro, C., & Perrotta, D. (Eds.). (2016). *Mobility and change in the Mediterranean area.* London/New York: Routledge.

Corrado, A., Palumbo, L., Caruso, F. S., lo Cascio, M., Nori, M., & Traindafyllidou, A. (2018). *Is Italian Agriculture a 'Pull Factor' for Irregular Migration—and, if so, why?* Open Society Foundations. https://www.opensocietyfoundations.org/sites/default/files/is-italian-agriculture-a-pull-factor-for-irregular-migration-20181205.pdf. Accessed 20 Jan 2019.

Dara Guccione, G., Ricciardi, G., & Guarnaccia, P. (2018). Immigrati e agricoltura: l'esperienza del progetto Sicilia Integra. In RRN (Rete Rurale Nazionale). *Terreni d'integrazione, 3*(31), 42–44.

De Lima, P., Jentsch, B., & Whelton, R. (2005). *Migrant workers in the Highlands and Islands.* Inverness: Highlands and Islands Enterprise.

Desjardins, M. R., Bessaoud, O., Issa, D., Berdaguer, D., Zied, A., Harbouze, R., & Debrun, A. (2016). *Une lecture de la crise migratoire: l'agriculture et le développement rural comme source de résilience dans les pays du Sud et de l'Est de la Méditerranée* (Watch Letter n 36). Montpellier: CIHEAM.

Donatiello, D., & Moiso, V. (2017). Titolari e riservisti. L'inclusione differenziale di lavoratori immigrati nella viticultura del Sud Piemonte. *Meridiana, 89*, 185–210.

Donatiello, D., & Moiso, V. (2018). Cooperazione, coordinamento, opportunismo. La filiera del Moscato d'Asti. *Meridiana, 30*, 135–154.

Donatiello, D., & Moiso, V. (2019). Rifugiati nell'imprenditorialità. Prospettive di sviluppo locale e percorsi di integrazione in un territorio DOCG. *Mondi Migranti* (Forthcoming).

ENRD. (2016). *Migrant and refugee integration.* Luxembourg: Publications Office of the European Union.

FAO. (2018). *The State of Food and Agriculture 2018. Migration, agriculture and rural development.* Rome.

Farinella, D., Nori, M., & Ragkos, A. (2017). Change in Euro-Mediterranean pastoralism: Which opportunities for rural development and generational renewal? In C. Porqueddu, A. Franca,

G. Molle, G. Peratoner, & A. Hokings (Eds.), *Grassland reources for extensive farming systems in marginal lands: Major drivers and future scenarios* (Grassland Science in Europe) (Vol. 22, pp. 23–36). Wageningen: Wageningen Academic Publishers.

Galera, G., et al. (2018). *Integration of migrants, refugees and asylum seekers in remote areas with declining populations*. (OECD Local Economic and Employment Development (LEED) Working Papers, 2018/03). Paris: OECD Publishing. https://doi.org/10.1787/84043b2a-en

Gallo, R. S., & Rioja, L. C. (2016). Inmigrantes, estrategias familiares y arraigo: las lecciones de la crisis de las areas Rurales. *Migraciones, 39*, 3–31. https://doi.org/10.14422/mig.i40y2016.00840services.

Gertel, J., & Sippel, R. S. (Eds.). (2014). *Seasonal workers in Mediterranean agriculture: The social costs of eating fresh*. London: Routledge.

http://nuevossenderos.es/

http://www.ka.org.tr/

Iocco, G., & Siegmann, K. A. (2017). A worker-driven way out of the crisis in Mediterranean agriculture. *Global Labour Column, 289*. http://column.global-labour-university.org/2017/09/a-worker-driven-way-out-of-crisis-in.html

Iocco, G., Lo Cascio, M., & Perrotta, D. (2017). *Mutualism, agriculture and migrant workers in southern Italy*. Paper presented at the international seminar "Crises of the models? Agricultures, territorial recompositions and new urban-rural relations", Tunis, 12–14 October 2017 [Unpublished].

Iocco, G., Lo Cascio, M., & Perrotta, D. (2018). Agriculture and migration in rural soutern Italy in the 2010s: New popolisms and new rural mutualism. *Conference paper n.77, ERPI 2918 International Conference*.

Iocco, G., Lo Cascio, M., & Perrotta, D. (2019). Lavoro migrante, mercati nidificati e sviluppo rurale nelle aree ad agricoltura intensiva del Sud Italia: due esperienze in Calabria e Sicilia. *Mondi Migranti, (3)*, 37–51. https://doi.org/10.3280/MM2019-001003.

Kasimis, C. (2008). Survival and expansion: migrants in Greek rural regions. *Population, Space and Place, 14*(6), 511–524. https://doi.org/10.1002/psp.513.

Kasimis, C. (2010). Demographic trends in rural Europe and migration to rural areas. *AgriRegioniEuropa, 6/21*. https://agriregionieuropa.univpm.it/it/content/article/31/21/demographic-trends-rural-europe-and-international-migration-rural-areas

Kasimis, C., & Papadopoulos, A. G. (2013). Rural transformations and family farming in contemporary Greece. In D. Ortiz-Miranda, E. V. Arnalte Alegre, & A. M. Moragues Faus (Eds.), *Agriculture in Mediterranean Europe: Between old and new paradigms research in rural sociology and development* (Vol. 19, pp. 263–293). Bingley: Emerald.

Kasimis, C., Papadopoulos, A. G., & Pappas, C. (2010). Gaining from rural migrants: Migrant employment strategies and socio-economic implications for rural labour markets. *Sociologia Ruralis, 50*(3), 258–276. https://doi.org/10.1111/j.1467-9523.2010.00515.x.

King, R., Lazaridis, G., & Tsardanidis, C. (Eds.). (2000). *Eldorado or fortress? Migration in Southern Europe*. London: Palgrave Macmillan.

Marsden, T., & Franklin, A. (2013). Replacing neoliberalism: Theoretical implications of the rise of local food movements. *Local Environment, 18*(5). https://doi.org/10.1080/13549839.2013.79715.

Mas Palacios, A., & Morén-Alegret, R. (2012). *International immigration around natural protected areas in Spain and Portugal* (XIII World Congress on Rural Sociology). Lisboa: IRSA.

Membretti, A., Kofler, I., & Viazzo, P. P. (Eds.). (2017). *Per forza o per scelta. L'immigrazione straniera nelle Alpi e negli Appennini*. Roma: Aracne.

Mostaccio, F. (2013). *La guerra delle arance*. Soveria Mannelli: Rubbetino.

Nori, M. (2015). Pastori a colori. *AgriRegioniEuropa 11/43*. http://tinyurl.com/zl7s9q2

Nori, M. (2017a). *Immigrant Shepherds in Southern Europe* (E-paper). Heinrich Böll Stiftung Foundation. https://www.boell.de/en/agriculture-food-production-and-labour-migration-southern-europe

Nori M. (2017b). Migrant Shepherds: Opportunities and Challenges for Mediterranean Pastoralism. *Journal of Alpine Research, 105*/4 https://rga.revues.org/3544

Nori, M., & López-i-Gelats, F. (2017). *Relevo generacional e inmigrantes en el mundo pastoril: el caso del Pirineo catalán.* Paper presented at the CSIC conference in Madrid.

Nori, M., & Luisi, D. (2019). Foreigners in Alpine to Apennine inner areas. The evidence of a territorial public policy. In G. Galera, I. Machold, A. Membretti, & M. Perlik (Eds.), *Alpine Refugees. Foreign immigration in the mountains of Austria, Italy and Switzerland.* Springer. https://www.cambridgescholars.com/alpine-refugees

Oliveri, F. (2015). A network of resistances against a multiple crisis. SOS Rosarno and the experimentation of socio-economic alternative models. *PArtecipazione&COnflitto, 8*(2), 504–529. http://siba-ese.unisalento.it/index.php/paco/article/view/15163/13200

Ortiz-Miranda, D., Moragues-Faus, A., & Arnalte-Alegre, E. (Eds.). (2013). Agriculture in Mediterranean Europe between old and new paradigms. In *Research in rural sociology and development* (Vol. 19). Bingley: Emerald.

Osti, G., & Ventura, F. (Eds.). (2012). *Vivere da stranieri in aree Fragili.* Napoli: Liguori.

Papadopoulos, A. G., & Fratsea, L. M. (2017). *Temporary migrant workers in Greek agriculture* (E-paper). Heinrich Böll Stiftung Foundation. https://www.boell.de/sites/default/files/e-paper_temporary-migrant-workers-in-greek-agriculture.pdf

Papageorgiou, F., Milnes, C., & Mylonas, D. (Eds.). (2016). *A capacity building manual for NGOS promoting the integration of migrants and refugees in rural areas.* Euracademy Association.

Perlik, M., & Membretti, A. (2018). Migration by necessity and by force to mountain areas: An opportunity for social innovation. *Mountain Research and Development, 38*(3), 250–264. https://doi.org/10.1659/MRD-JOURNAL-D-17-00070.1.

Perrotta, D., & Sacchetto, D. (2015). Migrant farmworkers in Southern Italy: Ghettoes, caporalato and collective action. *International Journal on Strikes and Social Conflicts, 1*(5), 75–98.

Ragkos, A., Koutsou, S., Theodoridis, A., Manousidis, T., & Lagka, V. (2018). Labor management strategies in facing the economic crisis. Evidence from Greek livestock farms. *New Medit Journal.* https://doi.org/10.30682/nm1801f.

Robinson, S., Hinojosa-Ojeda, R., & Thierfelder, K. (2017). *NAFTA and immigration: Linked labor markets and the impact of policy changes on the U.S. economy.* Washington, DC: Peterson Institute for International Economics and International Food Policy Research Institute.

Sarlo, A., & Martinelli F. (2016). *Housing and the social inclusion of immigrants in Calabria: The case of Riace and the 'dorsal of hospitality'* (COST Action IS1102. Working Paper No. 13). https://doi.org/10.12833/COSTIS1102WG2WP13.

Scholten, P., Baggerman, F., Dellouche, L., Kampen, V., Wolf, J., & Ypma, R. (2017). *Policy innovation in refugee integration? A comparative analysis of innovative strategies toward refugee integration in Europe* (Report for Dutch Department of Social Affairs). EMDI-IMISCOE.

Schrover, M., Van der Leun, J., & Quispel, C. (2007). Niches, labour market segregation, ethnicity and gender. *Journal of Ethnic and Migration Studies, 33*(4), 529–540. https://doi.org/10.1080/13691830701265404.

Semprebon, M., Marzorati, R., & Garrapa, A. M. (2017). Governing agricultural migrant workers as an "Emergency": Converging approaches in Northern and Southern Italian Rural Towns. *International Migration.* https://doi.org/10.1111/imig.12390.

SNAI. (2015). *Documenti del programma* (Strategia Nazionale Aree Interne). Roma: Ministero Sviluppo Economico.

Van der Ploeg, J. D. (2008). *The new peasantries. Struggles for autonomy and sustainability in an era of empire and globalization.* London: Earthscan.

Weidinger, T. (2018). *Residential mobility of refugees in rural areas of Southeastern Germany: Structural contexts as influencing factors.* Newcastle: Cambridge Scholars Publishings.

www.agricultures-migrations.org/en/
www.alpine-space.eu/projects/pluralps/en/home
www.cepaim.org
www.padima.org
www.risehub.org
www.sosrosarno.org
Zetter, R. (2017). Changing contexts, persistent challenges: the political and social milieu of refugee and asylum seeker reception in Europe. *Mondi Migranti, 3*, 7–13.

Chapter 5
Rural Origin Areas: Impacts and Practices

This chapter looks at the implications, impacts and consequences of rural migration on the areas of origin, where oftentimes portions of the family, and of the family assets, remain.

The chapter provides some basic indications concerning the triggers and impacts of migratory processes in areas of rural emigration in Northern Africa and Eastern Europe. It looks at the implications at the individual, household and community levels, as well as on the patterns of local development.

By the end of the chapter some elements and references for further and deeper analyses are proposed.

5.1 Introduction

In Chap. 3 we assessed the pull factors attracting migrants to EU agricultural settings, and in Chap. 4 we have addressed the implications of such immigration from the perspective of the areas of destination. This chapter looks at the reasons, the impacts and the consequences of rural migration in the immigrants' communities of origin, where oftentimes portions of the family, and of the family assets, remain.

Why do people leave their rural communities? What are the drivers and the triggers that induce such emigration? What are the implications for those remaining behind? What are the impacts on local communities and development patterns?

In the following sections we will partially answer these questions by looking at the main push factors, as well as at the local implications of rural emigration. Together with remittances, financial transfers sent by migrants back to their households, rural migratory flows carry implications for members who remain behind. As networks and relationships at the community and household levels are redefined, local socio-economic dynamics are also impacted, influencing patterns of local agriculture and rural development.

© The Author(s) 2020
M. Nori, D. Farinella, *Migration, Agriculture and Rural Development*, IMISCOE
Research Series, https://doi.org/10.1007/978-3-030-42863-1_5

The impact of emigration on origin communities is a huge domain with dedicated literature and debates. Here we present just a brief outlook of main relevant elements; reference bibliography for the EUMed context is also proposed for further and deeper analyses. Most data and information from the field are sourced from works in Tunisia with the RUMIT project (Zuccotti et al. 2018) and in Romania within the TRAMed project (Nori 2017; Nori et al. 2019).

5.2 Drivers of Emigration

Migration out of rural areas is triggered by various elements, usually defined as 'push factors', which include specific socio-cultural conditions, historical and geographical factors, economic development paths, political events, environmental and climatic factors. Somewhere rural emigration might also be conceived as an adaptive measure to ease the human pressure on a dwindling resource base and to alleviate rural unemployment, or as a response to the collapse of traditional organizational and/or governance mechanisms, including insecurity and conflict (de Haas 2009; Azzopardi 2012; UNDP 2015; Deotti and Estruch 2016; Desjardins et al. 2016; Zuccotti et al. 2018). Different drivers affect the intensity, trajectory and duration of rural migratory patterns (Nori et al. 2019).

Determinants of migration decisions can be analysed at three levels. The first level of analysis, called micro-level, focuses on individual migration decisions that are influenced by a migrant's individual features. The meso-level analysis looks at the socio-economic characteristics of the migrant's household of origin. Finally, the macro-level analysis focuses on the contextual features of the migrant's area of origin. These three levels of determinants do not exclude each other but can be considered complementary in explaining emigration. The Sustainable Livelihood approach (SLA) provides a consistent perspective to analyse these levels and the intertwined interactions (Frankenberger 2000; Ellis 2003; Scoones 2015).

One of the main reasons to emigrate is economic, and the youth leave rural areas and agriculture principally to look for better income, employment opportunities and for improving living conditions. This phenomenon could be indicative of failures of the institutional and/or market domains, whereby rural areas suffer from unemployment syndrome and agricultural income does not provide for a decent and sustainable livelihood (also refer to de Haas 2009; ILO 2015; Milan 2016; Gertel and Hexel 2018; Zuccotti et al. 2018).

More generally, agriculture and rural development provide little incentives for decent livelihoods, as these have in many countries been overlooked for decades in policy frameworks. Policies have been conceived to favour urban consumers rather than rural producers (ie. through food security strategies, subsidy schemes, pricing mechanisms, import schemes, etc.). This adds to a situation where risks are higher, and incomes lower compared to other economic sectors, with relevant implications for the economic and social viability of agriculture and rural livelihoods. The socio-cultural aspects of such marginalisation should not be underestimated; as the associated negative image agriculture has inherited vis-à-vis younger generations represents an important factor for their disengagement (Nori et al. 2019).

> **Box: Structural Challenges for Rural Maghreb and Mashreq**
> Poverty and unemployment in the Maghreb and Mashreq are concentrated in rural areas. Moreover most agricultural systems in the region are confronted with structural challenges: climate trends are putting under strain the limited water resources in arid regions; population growth, urbanization, erosion of soil fertility are reducing arable land; governmental policies favouring a liberalization process have facilitated a dual agricultural system dominated by highly competitive farms beside small, low-income family ones (World Bank 2011; CIRAD 2017).
> Additionally, the Southern region of the Mediterranean basin is characterized by a wide economic and social divide between inner, rural communities and urban areas, usually situated on the coast. These latter have access to the advantages of global economic exchange while the former are more isolated and face higher levels of poverty and unemployment, which are reflected in a higher vulnerability to food insecurity (CIHEAM 2009, 2014; Zuccotti et al. 2018).

The wider geo-political framework is also relevant to set the scene. Political turmoil, the recent financial crisis and related social and political events, including the Arab spring, international conflicts, terrorism threats and shifting EU border policy regimes have also had an influence on migratory patterns in the Mediterranean, and beyond.

> **Box: Emigration and Revolution in Rural Tunisia**
> In rural Tunisia outstanding differentiations have been reported between migratory projects that took place before 2011 and those that materialised afterwards in terms of opportunities, costs and impacts (Bardak 2014; Zuccotti et al. 2018).
> Figure 5.1 reports the reasons for migration before and after the 2011 political turmoil, from the viewpoint of the members of migrants' origin household (N = 633 men and 323 women. The results and the analysis come from the work undertaken in rural Tunisia in 2018 by the EUI Migration Policy Center (MPC) with the objective to enhance the understanding of rural out-migration by young people in Tunisia to facilitate positive impacts on food security, agriculture and development in rural areas (Zuccotti et al. 2018:42).
>
> There are important differences depending on whether the migrant is a man or a woman, as well as depending on whether migration happened before or in/after 2011. For men, work and improving life conditions are the two most important reasons for migration, and this applies to both periods of migration. Furthermore, sustaining the family, change in lifestyle and, to a lesser extent, study and reduction in income from agricultural activities, appear as more relevant reasons for migration among recent

(continued)

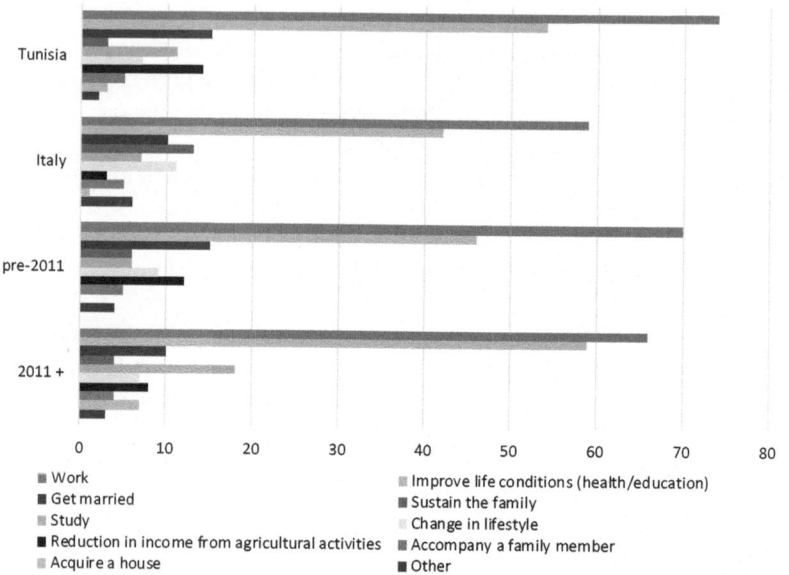

Fig. 5.1 Reasons for emigrating out of rural Tunisia. (Source: our-relaboration on data from Zuccotti et al. 2018. Map legend: N = 633 men and 323 women (multiple responses allowed))

migrants (i.e. with their first migration in 2011+). Among women the picture is quite different. Next to improving life conditions and work, getting married is an important reason for migration. (. . .) Migration for work and study is much more relevant for recent migrants than for migrants who left before 2011; conversely, migration to get married and to improve life conditions becomes less relevant among 2011+ migrants.

In Eastern Europe the fading of the Soviet regime has carried relevant consequences on the local socio-economic restructuring, with a specific impact on rural areas and societies. The process of integration into the European Union has provided a further trigger for the mobility of local populations out of rural areas.

Box: Moving off the Carpathian Mountains
After the collapse of the Soviet system, most Eastern European countries underwent quite complex socio-economic and political changes, with relevant impact on local rural livelihoods. The restructuring of rural livelihoods in areas such as the Carpathian mountainous were heavily impacted by several factors, including the shutting down of state-run factories, farms and mines at the end of the regime, the opening of frontiers and becoming EU citizens, money lost in wrong investments, recent curtails in public salaries and overall socio-economic disillusionment. The impacts of these factors differ according

(continued)

to diverse people and groups, as varieties existed amongst: those who worked for the State, those whose firm or enterprise collapsed, unemployed miners and workers, those that remained landless, those with family members abroad. Due to the current economic conditions of the country, many Romanians have tried coming back during the last decade, though several have and then decided to re-emigrate.

Recent literature on push factors in rural areas of origin reflects the growing concern in the scientific as well as policy circles about the structural drivers of emigration resulting from climate change, rural poverty and socio-political instability (refer amongst others to: Sivakumar et al. 2003; Vargas-Lundius and Lanly 2007; Lacroix 2009; Nori et al. 2009; de Haas 2010; Adams 2011; Scheffran et al. 2011; Hsiang and Burke 2014; Wodon et al. 2014; CIHEAM 2015).

The presence and evolution of existing migratory networks, set up by friends, relatives, or conationals who have undertaken emigration are also important factors that might contribute to influencing and facilitating decision-making on why, how and where to migrate, regardless of the initial drivers of migration (Bakewell et al. 2011; Mainwaring 2016). Especially in the case of seasonal workers, the recruitment of peers in the village community might be facilitated by their return home during non-peak labour periods at the winter season. Recruitment is often carried out directly by word of mouth between workers and between companies. Cases are reported where these networks present problems of intermediation, with sometimes exploitative mechanisms (De Haas 2007; Nori 2017).

5.3 Impacts of Rural Emigrations

The impacts of migration on the areas of origin are highly context-dependent, and often heterogeneous, and depend on the nature of the migratory project (ie. extent, duration, trajectory, which individuals, etc.. . .) and on local variables, including the local economy, the characteristic of the households, etc.. . . (De Haas 2007, 2008; Kriaa 2013; Deotti and Estruch 2016; Zuccotti et al. 2018).

A comprehensive analysis requires an approach that considers migrants as actors and factors of development also in their own countries of origin. When some members emigrate away from an area, a community or a household, the implications for those remaining and for local society can be assessed through three main domains (FAO 2018):

1. the *flow of migrants*, which affects the structure and composition of households and communities of origin, including labour supply, and could affect pressures on local resources and development patterns;
2. the *financial transfers, or remittances*, sent back by migrants to their households, which are often reinvested or spent locally;

3. the non-monetary transfers, often referred to as *"social remittances":* ideas, skills, technologies and social patterns brought or transmitted back by migrants and that expose origin communities to change, including potential impacts in furthering emigration.

These factors have controversial implications for the management of territories, the reproduction of community and household structures and for the reconfiguration of local class, gender and generational dynamics. Rural emigration may reduce pressures on local resources and increase community exposure to technical innovations and financial investments through the transfer of know-how and remittances, which could be reinvested locally. However, migration can also be problematic in terms of labour shortage or increased social disparities at community level.

As it is normally the active workforce that emigrates, this can be problematic for agriculture in terms of resource allocation and rural labour markets. For example emigration has direct consequences on local productivity, farm management and territorial maintenance, which might ultimately affect the resilience and the sustainability of local agrarian systems.

Existing territorial disparities and social inequalities that characterize rural contexts could be strengthened, as emigration might foster individualisation of resources, polarize production systems, abandonment of marginal territories and traditional know-how, economic differentiation and decreases in social protection measures (Bleahu 2004; King and Vullnetari 2006; Gertel and Breuer 2010; Amara and Jemmali 2016).

The impacts on community dynamics and the remaining members of households are important to understand. Some studies point to the psychological and emotional distress for those left behind due to the departure of a relative, usually the father or husband (Kriaa 2013; Zuccotti et al. 2018:38). As an example, the absence of parents from home might cause problems to educational and disciplinary experiences.

Furthermore, remaining members might have to take on the role of those who have emigrated, and thus take on new economic and social responsibilities, with important consequences of local gender relations. In some Eastern European countries where it is mostly rural women who emigrate to work as care-takers in the EU, remaining men have to reconsider their role within the household. While in the MENA, where it is often rural men emigrating, this leads to the feminization of agricultural labour, as women supply the labour missing from the departure of men.

Cases are reported whereby emigration could perpetuate gender imbalances, such as the reproduction of a patriarchal order, while other cases show that emigration challenges existing structures and dynamics, contributing to important social transformations. A brief review of the literature suggests that the impact of emigration and remittances on the participation of women to the labour market can be very heterogeneous, as these mechanisms do not operate in a socio-cultural vacuum and they need to be interpreted in light of the context specificities (David and Lenoël 2016).

Examples exist where with the departure of family men the social and economic empowerment of women have improved, increasing their level of autonomy and self-employment. Other cases occur where remittances have decreased the exposure of rural women to the formal labour market, thus somehow dis-empowering them

(Massey et al. 1993; Vargas-Lundius et al. 2008; Mahdi 2014; Lenoël 2014; Sampedro et Camarero 2015; Zuccotti et al. 2018).

Remittances are an important component of the migration project. Typically, the opportunity to send money back to the family is a main reason some members decide to emigrate. According to the International Fund for Agricultural Development (IFAD 2017) around 40% of international remittances globally are sent to rural areas.

In rural areas access to these revenues can make a great difference in local livelihood strategies. Evidence suggests that remittances from emigrated members have overtaken agriculture as the main source of income and investment, in a large number of rural communities (IFAD 2017; FAO 2018). Such financial sources have allowed rural households to expand, diversify or protect their livelihoods, with mostly positive outcomes associated with poverty reduction, food security as well as nutritional levels. Remittance money is largely utilised to satisfy the basic needs of the household, with a view to enhance basic living standards. Improved housing conditions, food consumption and access to basic services are often mentioned as main sources of remittance expenditure. For those who are abroad and send remittance, children education is often a priority, together with health care for elderly family members.

Figure 5.2 reports on the responses provided by migrants and by their origin households on the use they make of remittance. Remittances use patterns are more related to consumption and to improving the livelihoods in origin, rather than to new

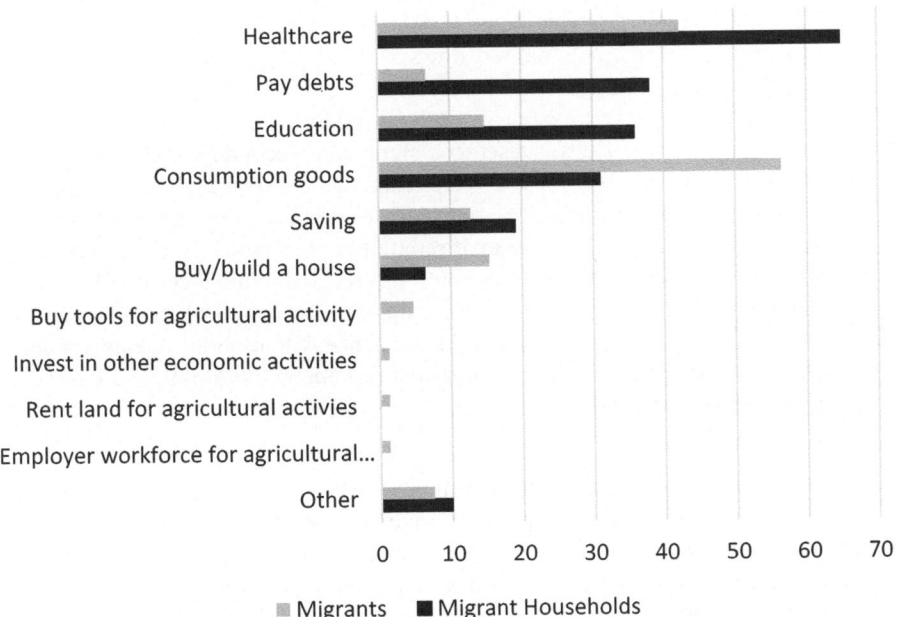

Fig. 5.2 Use of remittances. (Source: own elaboration of Zuccotti et al. 2018: 67. Map Legend: N = 413 origin migrant households; 195 migrants)

investments. Healthcare expenses, consumption of goods, education, paying debts and, to a lesser extent saving and buying a house, are the most common uses of remittances.

> **Box: Housing Development Through Remittance (Nori 2017)**
> In the Carpathian region of Romania remittances are typically used to purchase or build larger houses, or for aspects associated to prestige and 'social competition' such as expenses related to cars and clothes. Rural villages tend to be over-housed compared to the effective population, as most houses are built with money earned abroad. The large number of local houses on sale, '*de vanzere*', probably indicates the choice of the family not to come back, not even at a later stage.

> **Box: Remittances in Stateless Somalia (FSNAU 2013)**
> Somalia has seen its central state collapse in 1991. Since then, forms of local governance in the different Somali regions, and livelihood patterns, have evolved and adapted. The relevance of remittances sent by the international diaspora has been very important in shaping rural livelihoods in the Somali drylands.
>
> Remittances to the country were estimated at around US$1.2 billion per year in 2010. The relative significance of this amount can be appreciated when comparing it with international aid flows which averaged $834 million/year between 2007 and 2011, Foreign Direct Investment estimated at $102 million in 2011, and revenue from exports of $516 million in 2010.
>
> Evidence from the study attests to a significant secondary distribution of remittances, particularly by rural recipients; two rural households in three redistributed the remittance they received to support other rural relatives. Furthermore, households that receive remittances are more likely to support poorer relatives (75%) than those who do not receive remittances (54%).
>
> The relevance of remittance in supporting the livelihood of rural communities is thus dramatic, and it is likely that most needy households would not be able to cope with growing climatic and financial uncertainties if not supported by these remittance flows.

While money generated through agriculture is often reported as a main source of household funding for migratory projects, investing remittance in local agricultural assets and productive investments is less automatic than it might sound. Experience indicates that if any money is invested in agriculture, it is often in sub-sectors or activities that secure economic or social returns, through the purchase of land, livestock and agricultural equipment (ie. irrigation pumps, horticulture seeds), and the hiring of non-family labour force.

Investment strategies are tailored to the local agro-ecological or socio-cultural contexts, and might support agricultural intensification or enhance off-farm diversification of the rural economy, including service provision, agricultural diversification, value adding processing, agro-tourism, and environmental management. These investments might in turn attract workforce from neighboring regions and countries, thus contributing to perpetuating rural migratory flows.

> **Box: Land and Remittance in the Maghreb (Zuccotti et al. 2018)**
> Access to and control over land remains a main constraining factor to agriculture production and rural development in most Maghreb countries. As a result of emigration, many plots are abandoned or remain vacant, thus to affect agricultural productivity and the effective management of the natural resource base. Land reforms are debated in policy fora, while on the ground evolutions in contracts, transactions, rental and sale schemes of farming lands have evolved in recent times. Tree and livestock productions seem increasingly preferred for investment over annual crops, as they reportedly fit better within the socio-economic (ie. labour feminization, market demands) as well as agro-ecological (ie. dry spells, land fragmentation) dynamics.

Apart from business-oriented approaches that try exploiting existing lucrative opportunities, remittance reinvested in local rural development could address the need to maintain a social link with communities of origin (often with a view/dream to return at an elderly stage), to provide employment and income-generating opportunities to household members (ie. petty trading and local services provision), or to more broadly support the economic activities of local family members.

Problems around reinvesting in local agriculture development are often associated with high costs and limited returns, lack of local workforce, and limited market opportunities to sell products. For countries in the MENA, constraints related to accessing land, water, credit and commercial opportunities are indicated as main factors constraining local reinvestment of migrants' remittance. In eastern European countries the main reported constraints are associated with bureaucracy, limited market outlets and high production costs (for further references: Zuccotti et al. 2018; Nori et al. 2019).

Together with remittances, emigration can also contribute to transferring knowledge, skills and technologies that can play important roles in local development patterns. The implications and impacts of these 'social remittance' might provide further challenges to existing socio-cultural as well as economic patterns, including in social, gender and generational terms.

Here is a non-exhaustive list of the existing literature for some Mediterranean countries from which emigratory processes to southern Europe has been taking place in recent decades:

- For Romania: Bleahu and Janowski 2002; Bleahu 2004; Sandu 2005; Serbescu 2009; Cingolani 2009.
- For Albania: Eniel 2003; King and Vullnetari 2006; Miluka et al. 2010; Mendola and Carletto 2012.

- For Morocco: De Haas 2001; Mejjati Alami 2004; Azzarri et al. 2006; Khachani 2007; Gertel and Breuer 2010; Lenoël 2014; David and Lenoël 2016.
- For Tunisia: Boubakri 2013; Amara and Jemmali 2016; David and Marouani 2017.

Development interventions from international agencies could represent an important leverage to redress the institutional and structural aspects of local rural development so to become more attractive for remittance investments. Inclusive policy should consider the participation of government officials, local authorities, banks and microfinance institutions, trade unions and producers' associations, agricultural centres and diaspora networks (Nori et al. 2019).

References

Adams, R. H. (2011). Evaluating the economic impact of international remittances on developing countries using household surveys: A literature review. *Journal of Development Studies, 47*(6), 809–828. https://doi.org/10.1080/00220388.2011.563299.

Amara, M., & Jemmali, H. (2016). Deciphering the relationship between internal migration and regional disparities in Tunisia. *Social Indicators Research, 135*, 313–331. https://doi.org/10. 1007/s11205-016-1487-y.

Azzarri, C., Carletto, G., Cadvis, B., & Zezza, A. (2006). *Choosing to migrate or migrating to choose: Migration and labour choice in Albania* (ESA Working Paper No. 06–06). Rome: FAO.

Azzopardi, R. M. (2012). Recent international and domestic migration in the Maltese Archipelago: An economic review. *Island Studies Journal, 7*(1), 49–68.

Bakewell, O., de Haas, H., & Kubal, A. (2011). *Migration systems, pioneers and the role of agency*. Norface Research Programme on Migration, Department of Economics, University College London.

Bardak, U. (2014). *Labour market and education: Youth and unemployment in the spotlight*, in IEMed Mediterranean Yearbook 2014, pp. 78–84.

Bleahu, A. (2004). *Romanian migration to Spain motivation, networks and strategies*. Bucarest: Institute for Quality of Life, Romanian Academy.

Bleahu, A., & Janowski, M. (2002). *Rural non-farm livelihood activities in Romania: A report on qualitative fieldwork in two communities* (NRI Report 2725). Chatham: Natural Resources Institute.

Boubakri H. (2013). *Revolution and International Migration in Tunisia* (MPC Research Reports 2013/04, Robert Schuman Centre for Advanced Studies). San Domenico di Fiesole (FI): European University Institute.

CIHEAM. (2009). Mediterra, rethinking rural development in the Mediterranean. In B. Hervieu & H. L. Thibault (Eds.), *International centre for advanced Mediterranean agronomic studies and blue plan*. Paris: Presses de Sciences-Politiques.

CIHEAM. (2014). *Mediterra, logistics and agro-food trade, a challenge for the Mediterranean* (A CIHEAM Report). Paris: Presses de Sciences-Politiques.

CIHEAM. (2015). Crise et résilience en la Méditerranée. *Watch Letter 36*. Montpellier: Centre international de hautes études agronomiques méditerranéennes. http://www.iamb.it/share/integra_files_lib/files/WL36.pdf

Cingolani, P. (2009). *Romeni d'Italia. Migrazioni, vita quotidiana e legami transnazionali*. Bologna: il Mulino.

CIRAD, CIHEAM-IAMM. (2017). *L'agriculture familiale á petite échelle au Proche Orient et Afrique du Nord*. Montpellier.

David, A., & Lenoël, A. (2016). International Emigration and the Labour Market Outcomes of Women Staying Behind – The Case of Morocco. Papiers de Recherche AFD, n°2016–23.

David, A., & Marouani, M. A. (2017). *Migration patterns and labor market outcomes in Tunisia* (No. DT/2017/03).

De Haas, H. (2001). *Migration and agricultural transformations in the oases of Morocco and Tunisia*. Utrecht: KNAG.

De Haas, H. (2007). *Remittances, migration and social development: A conceptual review of the literature*, Social Policy and Development Programme Paper, n.34, United Nations Research Institute for Social Development.

De Haas, H. (2008). *Migration and development: A theoretical perspective* (Working Paper, Vol. 9). Oxford: International Migration Institute, University of Oxford.

De Haas, H. (2009). International migration and regional development in Morocco: A review. *Journal of Ethnic and Migration Studies, 35*, 1571–1593. https://doi.org/10.1080/13691830903165808.

De Haas, H. (2010). Migration and development: A theoretical perspective. *International Migration Review, 44*(1), 227–264. https://doi.org/10.1111/j.1747-7379.2009.00804.x.

Deotti, L., & Estruch, E. (2016). *Addressing rural youth migration at its root causes: A conceptual framework*. Rome: FAO.

Desjardins, M. R., Bessaoud, O., Issa, D., Berdaguer, D., Zied, A., Harbouze, R., & Debrun, A. (2016). *Une lecture de la crise migratoire: l'agriculture et le développement rural comme source de résilience dans les pays du Sud et de l'Est de la Méditerranée* (Watch Letter n°36). Montpellier: CIHEAM.

Ellis, F. (2003). *A livelihoods approach to migration and poverty reduction*. Norwich: Department for International Development (DFID) (No.CNTR- 034890).

Eniel, N. (2003). *Differenziazioni territoriali in Albania dalla caduta del comunismo*. Collana Tesi on-line AlessandroBartola Studi e ricerche di economia e di politica agraria. AgriRegioniEuropa.

FAO. (2018). *The State of Food and Agriculture 2018. Migration, agriculture and rural development*. Rome.

Frankenberger, T. R. (2000). A brief overview of sustainable livelihoods approaches. In *Proceeding from the forum on operationalizing sustainable livelihoods approaches, Proceedings*, FAO.

FSNAU. (2013). *Family ties: Remittances and livelihoods support in Puntland and Somaliland*. Nairobi: UN.

Gertel, J., & Breuer, I. (Ed.) (2010). *Pastoral Morocco: Globalizing scapes of mobility and insecurity*. University of Leipzig. Reichert Pubbl. Nomaden und Sesschafte collection.

Gertel, J., & Hexel, R., (Ed.) (2018). *Coping with uncertainty*. Youth in the Middle East and North Africa. Saqi books.

Hsiang, S. M., & Burke, M. (2014). Climate, conflict and social stability: What does the evidence say? *Climatic Change, 123*, 39–55. https://doi.org/10.1007/s10584-013-0868-3.

IFAD. (2017). *Sending money home: Contributing to the SDGs, one family at a time*. Rome: International Fund for Agricultural Development (IFAD).

ILO. (2015). *Global estimates of migrant workers and migrant domestic workers: Results and methodology*. Geneva: International Labour Office.

Khachani, M. (Ed.). (2007). *L'impact de la migration sur la société marocaine*. Rabat: AMERM—GTZ—Goethe Institute.

King, R., & Vullnetari, J. (2006). Orphan pensioners and migrating grandparents: The impact of mass migration on older people in rural Albania. *Ageing and Society, 26*(5), 783–816. https://doi.org/10.1017/S0144686X06005125.

Kriaa, M. (2013). *Etude de l'impact de la migration sur les familles de migrants présentes au pays*. UNFPA, UNICEF, IOM, OTE, Statistiques Tunisie and Confédération Suisse.

Lacroix, T. (2009). *Migration, Développement, Codéveloppement: quels acteurs pour quels discours?, Rapport de synthèse européen. Informer sur les migrations et le développement.* Paris: Institut Panos.

Lenoël, A. (2014). *Burden or empowerment? The impact of migration and remittances on women left behind in Morocco.* PhD Social Policy, University of Bristol.

Mahdi, M. (2014). L'émigration des pasteurs nomades en Europe: Entre espoir et désillusion. In J. Gertel & R. S. Sippel (Eds.), *Seasonal workers in Mediterranean agriculture: The social costs of eating fresh.* London: Routledge.

Mainwaring, C. (2016). Migrant agency: Negotiating borders and migration controls. *Migration Studies, 4*(3), 289–308.

Massey, D., Douglas, S., Arango, J., Graeme, H., Kouaouci, A., Pellegrino, A., & Edward Taylor, J. (1993). Theories of international migration: A review and appraisal. *Population and Development Review, 19*(3), 431–466.

Mejjati Alami, R. (2004). Femmes et marché du travail au Maroc. *L'Année du Maghreb, 1,* 287–301.

Mendola, M., & Carletto, G. (2012). Migration and gender differences in the home labour market: Evidence from Albania. *Labour Economics, 19,* 870–880. https://doi.org/10.1016/j.labeco.2012.08.009.

Milan, A. (2016). *Rural livelihoods, location and vulnerable environments: Approaches to migration in mountain areas of Latin America.* Dissertation to obtain the degree of Doctor at the Maastricht University.

Miluka, J., Carletto, G., Davis, B., & Zezza, A. (2010). *The vanishing farms? The impact of international migration on Albanian family farming.* Washington, DC: World Bank. https://openknowledge.worldbank.org/handle/10986/4777

Nori, M. (2017). *Immigrant shepherds in southern Europe* (E-paper). Heinrich Böll Stiftung Foundation. https://www.boell.de/en/agriculture-food-production-and-labour-migration-southern-europe

Nori, M., El Mourid, M., & Nefzaoui, A. (2009). *Herding in a shifting Mediterranean: Changing agro-pastoral livelihoods in the Mashreq and Maghreb region.* Robert Schuman Centre, European University Institute, Florence. http://econpapers.repec.org/paper/erpeuirsc/p0223.htm

Nori, M., Triandafyllidou, A., Le Hénaff, M. H., Robert, C., Castro, G., Abdelali-Martini, M., & Provenzano, G. (2019). Forum on agriculture, rural development and migration in the Mediterranean. A better understanding of the drivers and impacts for forward-looking policies and programmes. In *Proceedings of a conference EUI, CIHEAM, UfM.* Firenze: European University Institute.

Sampedro, R., & Camarero, L. (2015). *International immigrants in rural areas: The effect of the crisis in settlement patterns and family strategies.* In *Proceedings of the XXVI Congress of the European Society for Rural Sociology,* Aberdeen (UK).

Sandu, D. (2005). Emerging transnational migration from Romanian villages. *Current Sociology, 53*(4), 55–82. https://doi.org/10.1177/0011392105052715.

Scheffran, J., Marmer, E., & Sow, P. (2011). Migration as a contribution to resilience and innovation in climate adaptation. *Applied Geography, 33*(1), 119–127. https://doi.org/10.1016/j.apgeog.2011.10.002.

Scoones, I. (2015). *Sustainable rural livelihoods and rural development.* Winnipeg: Practical Action Publishing/Fernwood Publishing.

Serbescu, A. (2009). On change and adaptation: Rural inhabitation during the Romanian post-socialist transition. *Traditional Dwellings and Settlements Review, 21*(1), 37–50. http://www.jstor.org/stable/41758711.

Sivakumar, M. V. K., Lal, R., Selvaraju, R., & Hamdan, R. (Eds.). (2003). *Climate change and food security in West Asia and North Africa.* Amsterdam: Springer.

UNDP. (2015). *Human development report.* New York: Palgrave Macmillan.

Vargas-Lundius, R., & Lanly, G. (2007). *Migration and rural employment.* Paper prepared for the round table organized by the policy division during the 30th session of the Governing Council of IFAD, Rome.

Vargas-Lundius, R., Villareal, M., Lanly, G., & Osorio, M. (2008). *International migration, remittances and rural development.* Rome: IFAD and FAO.

Wodon, Q., Liverani, A., Joseph, G., & Bougnoux, N. (Eds.). (2014). *Climate change and migration: Evidence from the Middle East and North Africa.* Washington, DC: World Bank Publications.

World Bank. (2011). *Migration and remittances. Factbook 2011.* Washington, DC: World Bank.

Zuccotti, C. V., Geddes, A. P., Bacchi, A., Nori, M., & Stojanov, R. (2018). *Rural migration in Tunisia. Drivers and patterns of rural youth migration and its impact on food security and rural livelihoods in Tunisia.* Rome: Food and Agriculture Organization of the United Nations.

Chapter 6
Mobility and Migration in Mediterranean Europe: The Case of Agro-pastoralism

While most of the existing literature on rural migrations focuses on immigrant workers in intensive agricultural systems, this work tries filling existing gaps by addressing more marginal and remote rural settings. The mountainous, inner and island territories that cover a large part of the Mediterranean are particularly affected by intense demographic decline, land abandonment and problems of generational renewal, posing important questions concerning the sustainability of local development.

The agro-pastoral systems that characterize these settings provide a relevant observatory to explore and disentangle rural migratory dynamics. In these areas the presence of immigrant shepherds represents a critical asset to maintain local farms, villages and territories alive and productive. By analysing the reconfiguration of human and natural landscapes in agro-pastoral systems of Greece, Spain and Italy, we provide a framework to assess the contribution of immigrants in maintaining and reproducing European rural societies.

6.1   Introduction

With a view to disentangle the contributions of immigrant communities to rural development, in this chapter we assess in deeper detail the relevance of foreign workers in agro-pastoralism.

Historically, agro-pastoralism – the extensive breeding of mostly sheep, goats and cattle associated to farming activities – is a traditional form of territorial management in marginal area of Mediterranean Europe. It provides valuable products and precious socio-ecosystem services, which are an integral part of food production and natural resource systems.

Shepherding, its practices and symbols are a traditional feature of all Mediterranean cultures; transhumance routes cut across EUMed territories creating synergies and enhancing economic integration; the wool economy has been a main driver of

© The Author(s) 2020
M. Nori, D. Farinella, *Migration, Agriculture and Rural Development*, IMISCOE
Research Series, https://doi.org/10.1007/978-3-030-42863-1_6

regional development for centuries. Animal proteins in the form of meat and milk products have become more relevant in recent decades, associated to several ecosystem services that are being increasingly acknowledged; most natural parks and reserves in the region are superimposed with historical agro-pastoral settings.

Agro-pastoralism is central especially in marginal rural settings, which constitute about half of the Mediterranean territories. Such territories are associated with lower income and employment opportunities, as well as with limited access to social, cultural and institutional services, pushing younger generations to move away in search of better options. Intense demographic decline, land abandonment and generational renewal represent important questions for the future of sustainable rural development.

Paradoxically, as it will be discussed, agro-pastoralism is increasingly appreciated by the European society for its products and services even though it is decreasingly practiced by its citizens. This paradox is currently resolved through the growing presence and contribution of immigrant shepherds. As farm labour intensifies and socio-economic conditions have hardly improved, foreign workers have become strategic for the survival of agro-pastoral enterprises, and for the marginal and depopulated territories.

By analysing the reconfiguration of human and natural landscapes in the agro-pastoral systems of Spain, Greece and Italy, we provide a framework to assess the contribution of immigrants in maintaining and reproducing European rural societies. Data, estimates and sources are specifically for sheep and goat farming, and for markets related to ewe's milk and sheep cheese, as these are typically associated to agro-pastoral systems in the region.

6.2   Contemporary Changes in Mediterranean EU Agro-pastoralism

To understand the increasing role of immigrant workers in agro-pastoralism, it is necessary to introduce the characteristics of agro-pastoral systems and their recent changes and dynamics.

Historically, Mediterranean rural settings are characterised by agro-pastoral systems. Due to the climatic and territorial features of the region, Mediterranean agricultural systems were typically extensive and oriented to self-consumption, muntifunctionality and polyculture; these consist of cereals such as wheat and barley, while vineyards, olive and other fruit trees are part of the countryside, especially where the steepest terrain makes it difficult to farming cereals. Associated to these crops, the extensive breeding of mostly sheep, goats and cattle represents a typical complementary component of local rural livelihoods (Campbell 1964; Pernet and Lenclud 1977; Le Lannou 1979; Ravis-Giordani 1983; Meloni 1984; Mattone and Simbula 2011).

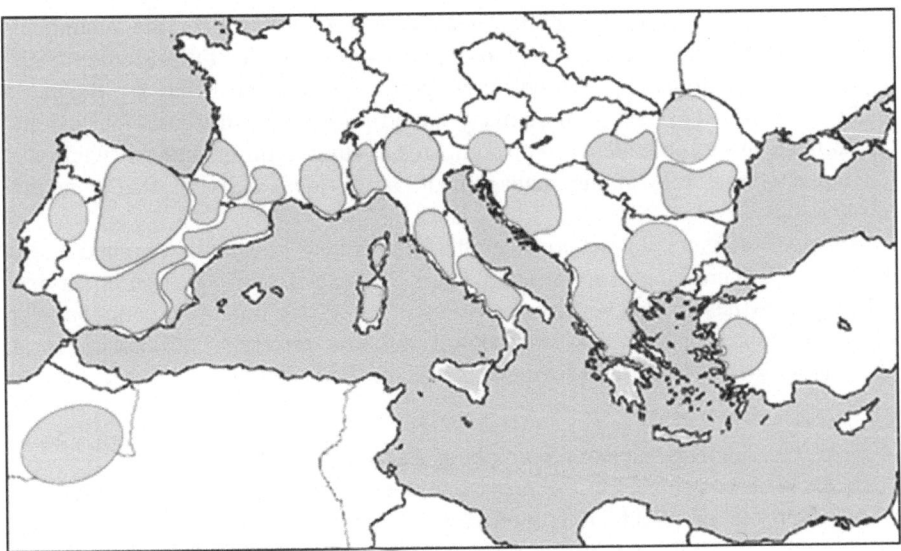

Fig. 6.1 Areas of main transhumant systems in the Mediterranean. (Source: Own elaboration from Braudel (1982))

Box: Mediterranean Transhumances

In the Mediterranean, breeding sheep and goats is often associated with the practice of transhumance, the seasonal mobility of flocks, which makes it possible to adapt flocks' productive and reproductive performances to the rhythm of the seasons and the availability of pasture – cooler inner, mountain pastures during the summer and milder coastal areas or valley bottoms in winter times. In complementarity with sedentary agricultural activities, this system enables making the best use of the agro-ecological diversity and of the marked seasonality of the region. Transhumance has been instrumental for the management and the governance of extended territories. For example, the Mesta systems in Spain, and the Dogana in Italy served the lucrative trade of wool, while also contributing to integrating territories, economies and cultures (Braudel 1982 – also refer to Fig. 6.1). During the last century, these systems began to fade due to important changes in global economy, regional trade and local land use trough the diversification of fibre markets. As it can be guessed living and working conditions associated to this practice are quite hard and require high degrees of rusticity and commitment.

Agro-pastoral systems are based on specific agricultural practice whereby animals are mostly raised in open settings, mostly feeding on local natural grazing; several interactions and synergies characterise the combination of farming and livestock systems.

Agro-pastoralism has proved to be an effective land use for the mountains, drylands, and islands that cover approximately half of the Euro-Mediterranean region. As a main source of food, employment and income it has represented a resilient livelihood system in these marginal territories where the costs for land and labour make this a convenient option compared to other forms of land use, while also playing a critical role in the management of the local rich but fragile natural resource base.

Agro-pastoralism is in fact considered a High Nature Value (HNV) practice, and it is as such increasingly appreciated for the so-called socio-ecosystem services (SES) it provides to the wider society, as it associates quality production with socio-economic opportunities and natural resource protection (Caballero et al. 2009; Nori 2015; Meloni and Farinella 2015a, b; IFAD 2017).

Box: The Socio-Ecosystem Services of Agro-pastoralism

Pastoralism contributes to the provision of ecosystem services, as it plays a relevant role in maintaining biodiversity. Apart from flora species, agro-pastoral farmers also contribute to the protection of diversity by rearing autochthonous animal breeds, supporting wildlife habitats and contributing to preserving landscapes. In addition, agro-pastoralism enhances the resilience to hydro-geological risks and natural hazards. Well-grazed vegetation represents an important factor to control erosion, flooding and landslides, as well as a barrier to the spread of forest fires, while enhancing the maintenance of biodiversity and soil quality and reduces the fire risk in permanent crops such as olive groves. The capacities of properly managed pasturelands to absorb carbon and water provide as well a most effective way to store CO_2 and to manage rainfall, two ecosystem functions that are increasingly important in a climate perspective (Caballero et al. 2009; Nori and de Marchi 2015; Moreira et al. 2016; Ragkos and Nori 2016).

Apart from economic and ecological aspects, agro-pastoralism in the Mediterranean plays an important socio-cultural and political role as well. By supporting local livelihoods it ensures human presence is maintained in harsh terrains and remote communities, thus contributing to averting socio-economic desertification, with relevant implications on the cultural heritage and territorial identity of local communities. This applies specifically in mountainous areas, 'Europe's ecological backbone', as much as in most EUMed islands, such as Sardinia, Crete, Majorca, Corsica, Peg, Cyprus, etc. Table 6.1 reports on the relevance of agro-pastoralism for the economy and society of Sardinia and Crete.

Although with many differentiations and local characterisations, the embeddedness of agro-pastoralism in local territories and societies contributed in time to shape and maintain the extraordinary cultural and biological diversity that characterizes the Mediterranean countryside, an "historical rural landscapes"

Table 6.1 Main features of agro-pastoral systems in Crete and Sardinia (year 2016)

Crete		Sardinia
0.630 Mils	*Human population*	1.663 Mils
1,1 Mils	*Sheep population*	3.15 Mils
4.800	*Sheep farms*	11.213
229	*Average size – heads/flock*	280
13% flocks	*Transhumance*	Few hundreds
Autochthonous breeds	*Breed*	Autochthonous Sarda breed
Multifunctional	*Production focus*	Mainly Pecorino Romano cheese
25% at local *mitata* level	*Dairy processing*	Mostly through industries
Mostly through cooperatives	*Marketing*	32 coops exist – but mostly controlled by the industry
4 PDOs	*Cheese market certification*	3 PDOs
30%, mostly from Albania, Bulgaria	*Proportion of immigrants amongst salaried shepherds*	20% mostly from Romania

Source: Nori et al. (2017)

(Antrop 1997, 2005; Agnoletti 2013). The complex articulation between pastoral resource management and natural ecosystems is well reflected in Mediterranean landscapes of high natural and cultural value such as the Causses and Cévennes, Dolomites, Picos de Europa, Parco Nazionale degli Abruzzi, Atlas mountains, etc.... These offer important opportunities for leisure as well as for tourism development in these regions.

> **Box: Capitalising on Agro-pastoral Traditions**
> A series of projects and initiatives have been launched to enhance pastoral cultures and contribute to their capitalization; these include the Virtual Museum of Transumanzia in Slovenia, the Museum of Transhumancia in Guadalajara and in Aigüestortes in Spain, la Maison du Berger in Provence, the Ecomuseum della Pastorizia in Val Stura and in Sardinia, and other Transhumance musea in the Abruzzi and Molise. The French park of Causses et Cévennes has been acknowledge by UNESCO as a world heritage for Mediterranean agro-pastoral cultural landscape, specifically characterised by pastoral resource management through extensive animal keeping.

These contributions of agro-pastoralism to sustainable societal development are recognized and supported by the Common Agricultural Policy (CAP). CAP plays today a significant role in the budget of agro-pastoral farms. On average CAP financial support might represent today half of the EUMed breeders' annual revenue, with trends and variations changing from a country to another depending on local legislations and implementation of CAP schemes (Nori 2015; Fréve 2014; Ragkos et al. 2016b).

It is usual to hear amongst agro-pastoral farmers that "we spend today more time in the office than in the field", or to read that "we are considered as landscape gardeners rather than producers of meat and milk" (Nori 2017b: 14) these financial contributions represent a critical resource for this sector (also refer to Brisebarre 2007; Nadal et al. 2010; Pitzalis and Zerilli 2013; Nori 2015). Without this support, sheep, goats and cattle would have already disappeared from most landscapes.

Box: The CAP vis-à-vis Agro-pastoralism

The recent reforms of the Common Agricultural Policy (CAP) have shifted the focus of public support and rural welfare towards a multifunctional vision of agriculture. In such context agro-pastoralists are increasingly demanded to play their role in managing natural resources and maintaining landscapes, while also contributing to stabilize population and to enhance socio-economic development in marginal settings (Nori and Gemini 2011; Beaufoy and Ruiz-Mirazo 2013). Following changing societal demands, CAP policy support has shifted through time from conceiving agro-pastoralists as mostly livestock producers to 'guardians of nature' or suppliers of multifunctional goods and socio-ecosystem services (Marsden 1995; McNally 2001; Vaccaro and Beltran 2007; López-i-Gelat 2013).

The role of the European Union and its CAP remains though contradictory and controversial. The ways CAP addresses the specific problems and needs of agro-pastoralists is not deemed adequate, as for several factors they fall within the same criteria with conventional intensive systems, or not recognised in their diversity – such as for grazing systems in forest areas not accepted for agro-ecological payments; or for the local processing of dairy products, which may at times not comply with European legislations regarding quality standards (e.g. cheese from raw milk), thus hindering the expansion of informal marketing networks and affecting their economic viability (Farinella et al. 2017; Farinella 2018).

Although this public policy is essential to keep these territories populated and productive, the constant decrease in agro-pastoral farms and operators seems to attest that CAP schemes do not seem to be an adequate guarantee for the permanence and the reproduction of these systems.

In this sense, as we will see, the use of low-cost migrant labor is often applied as a strategy to cut down on production costs to tackle the agricultural squeeze.

6.3 The Recent Dynamics of Agro-pastoralism in Greece, Spain and Italy

In the traditional agro-pastoral system the family has typically represented the pillar of farm management, with work organized according to gender and age and aimed at the production and reproduction of the household as well as of the herd. Within a domestic household economy traditional agro-pastoral system is naturally multifunctional in providing a variety of products and services.

As explained in chapter two for the wider agrarian world, the traditional agro-pastoral model has undergone intense restructuring and reconfiguration in recent decades. Agricultural modernization has led to the expansion of monoculture in lowland areas and the abandonment of several inner and marginal rural settings, less suitable for intensive agriculture. Changes in commercial practices, agricultural policies, societal attitudes and consumption habits have all contributed to transforming not only the agricultural economy, but rural society as a whole. Today most of the livestock products demanded by consumers are increasingly supplied by more intensive production systems, while typical pastoral products such as lambs, ewes' milk and goat cheeses are sourced through imports from other regions (Kerven and Behnke 2011).

EUMed countries (including Portugal and Mediterranean France) hosted in 2015, 39% and 67% of the EU-28 sheep and goat population respectively; these were also the highest producers of ewe's milk: first Greece (30.5% of total milk production in EU); Spain second (27.1%), and Italy third (21.3%) (ISMEA 2017). The sheep milk sector will be analysed through the following tables to assess the changes these systems have undergone in recent decades (Figs. 6.2–6.5).

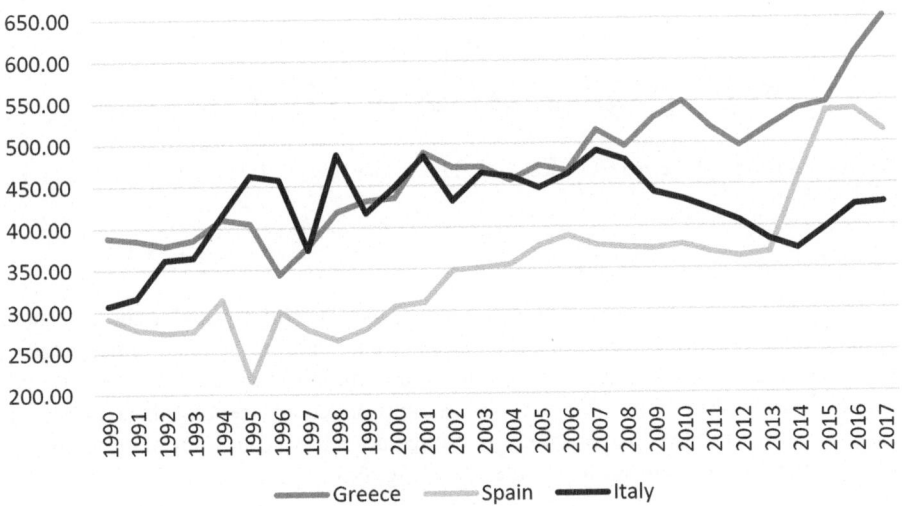

Fig. 6.2 Trend of sheep milk delivered to dairies in Greece, Spain and Italy. Products obtained (1000 t), years 1990–2017. (Source: Our elaboration on EuroStat data)

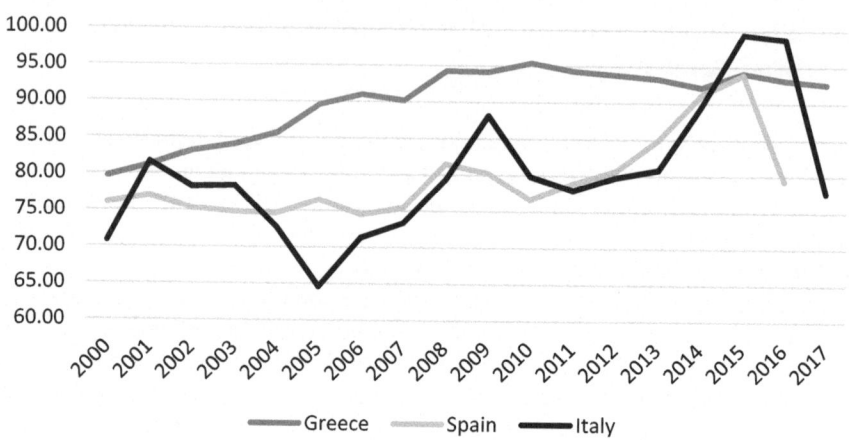

Fig. 6.3 Trends for sheep milk price in Greece, Spain and Italy – (€/Kg), years 2000–2017. (Source: Our elaboration on EuroStat data for Greece and Spain, ISMEA for Italy)

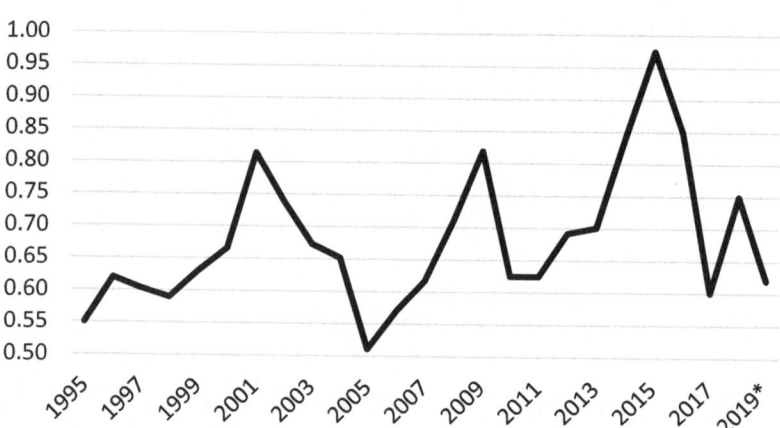

Fig. 6.4 Volatility of the price of sheep milk in Sardinia (€/liter). (Source: Our elaboration on ISMEA data)

The sheep-milk is processed in dairy industries to produce popular cheeses such as the Italian "Pecorino Romano", the Greek "Feta" and the Spanish "Manchego". The fact that these agro-pastoral products are mainly targeted to international markets and within large global food distribution chains represents a factor of economic vulnerability, as global commodities are exposed to high price volatility which affects many agro-pastoral farms (see Fig. 6.3).

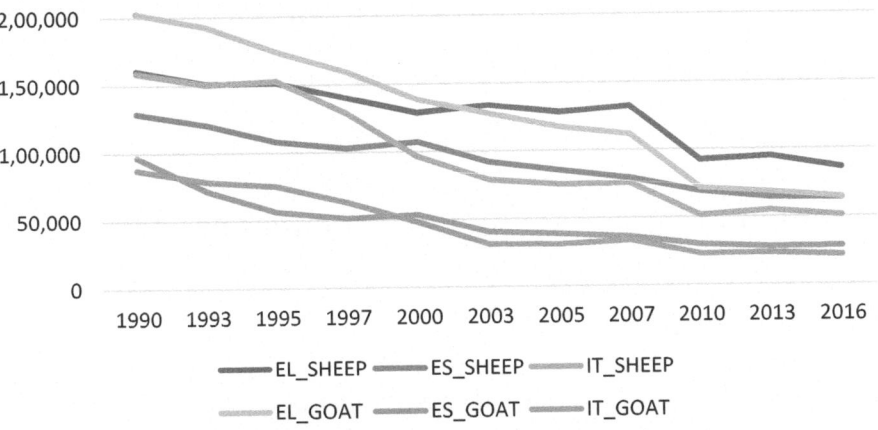

Fig. 6.5 Trends in sheep and goat farms in Greece, Spain and Italy (years 1990–2016). (Map legend: *EL* Greece, *ES* Spain, *IT* Italy. Source: Our elaboration on EuroStat data)

Box: Sardinian Agro-pastoralists Hanging on Global Value-Chains (Farinella 2018, 2019a)

Sardinia is an Italian region highly specialized in a sheep-breeding that is largely dependent on international markets. According to Istat data, in 2016 Sardinia holds 45.5% of the national sheep flock (3.3 million sheep) and produces 68.4% of Italian ewe milk production (2.9 million liters). With the highest ageing index and the lowest birth rate in the world, Sardinia is a region with a depressive demographic dynamic. In 2013, out of a total of 30.260 sheep farmers, over a third were aged above 60 and over 50% were older than 50, while only 5% were aged less than 30 years (Farinella et al. 2017).

Sardinian agro-pastoralism remains semi-extensive, though it is heavily embedded in the "Pecorino Romano" (PR) dairy value chain, a low value-added commodity based on large-scale production for export to the global market, mainly in the U.S.A. (which imports about 80% of the total PR production) where it is used as "mixed cheese" to improve the taste of industrial food. This makes Sardinian farmers largely dependent on the price of sheep milk, determined in turn by the value of PR in the global market. The PR supply chain is organized in an oligopolistic fashion, with the large processing industries and large-scale distribution brands controlling its sales and imposing the milk purchase price on sheep farmers. Strong fluctuations in the price of sheep milk can be observed, with a periodic and increasingly close frequency of price spikes, which trigger Sardinian livestock farmers to increase the levels of farm exploitation. It is within this vicious circle, that the costs of the market are dumped on the weakest links in the chain.

In 2015, the price of sheep raw milk was around 1 €/liter in Sardinia; since 2016 onwards it has gone through important fluctuations, with peaks of 0.60–0.50 €/liter, which eventually pushed Sardinian sheep farmers into unusual forms of protest in february 2019 (when the milk price is lowered to 0,60 €/liter), such as the pouring of milk on the streets (Simula 2019; Farinella 2019b). The Sardinian case is emblematic of the difficulties faced by farmers in global agri-food chain, which they often participate from a subordinate position.

Table 6.2 The consistency of the sheep sector in EUMed countries

Country	Sheep farms[a] (2016)	Sheep flock[b] (2016)	Ewes' milk – products (Kt)[a] (2017)	% meat[b] production (2010)	% milk[b] production (2010)
Greece	86.030	8.227.630	427.43	15%	85%
Spain	63.730	15.862.160	514.20	82%	18%
Italy	50.650	7.026.540	650.90	35%	65%
EUMed	200.410	31.116.330			

Sources: [a]EuroStat, [b]ISTAT, INE, Magrama

Due to these pressures from the institutional and market spheres, the costs related to increasing animal productivity have arisen consistently in latest decades. Further dependence on genetics, agronomic and veterinary sciences as well as through the increasing reliance on market-supplied inputs and intensified animal feed production have also amplified input costs (Farinella 2018, 2019a). However, most agro-pastoral practices continue to rely on physical labour and manual activity, which is poorly mechanized, and with low productivity compared to other agricultural systems. Productivity rates have often increased more slowly than production costs, which has contributed to increasingly squeeze the sector viability; many farmers have been either forced to close or to deeply restructure their farm, by expanding their herd and re-organising land and labour resources accordingly as a way to adjust cost-benefit ratios (Hadjigeorgiou 2011; Farinella and Meloni 2013; Mattalia et al. 2018; Farinella 2018, 2019a; Theodoridis and Ragkos 2018).

The exploitation of immigrant workers represents for these farms a strategy to contain costs and it is often accompanied by the self-exploitation of the farmers' OWN family work. Different from crop-oriented farms, agro-pastoralists work closely with their labourers and carry out tasks together, with more horizontal and less hierarchical relationships.

The restructuring of the sector following these dynamics has profoundly changed the size of agro-pastoral enterprises and the nature of livestock management. Today there is a marked separation between the managerial and the field work; on the one hand the burdensome administrative components to be compliant with technical requirements and financial assistance, and on the other hand the tending of the livestock.

The classic refrain amongst agro-pastoralists is that "20 years ago with a flock half size of the present one we had a decent life and we could even make savings and investments. Now with a double-sized flock, it is difficult to make ends meet by the end of the year" (Nori 2017b:12). Official data seem to confirm this perception (refer to Tables 6.3, 6.4, 6.5, and 6.6 and to Figs. 6.5, 6.6, 6.7, and 6.8). The amount of agro-pastoral farms and flocks have decreased (Table 6.3), while the size of those remaining in the business has expanded (Tables 6.4 and 6.6). The decline in agro-pastoral farms accompanies the marked overall reduction of about 30% of the EUMed small ruminants' flock in recent decades (FAO database; EuroStat 2016).

Table 6.3 Variations of sheep and goat farming data and rates in Greece (years 1990–2016)

Greece	1990	1995	2000	2005	2010	2016	% var. 1990–2016	% var. 2000–16
Farms with livestock			392.960	407.510	273.160	238.520		−39.3
Sheep farms	160.420	151.220	128.550	127.940	91.930	86.020	−46.4	−33.1
Sheep flock	8.258.130	8.328.130	8.752.670	9.066.370	9.156.820	8.227.630	−0.4	−6.0
Sheep/farm rate	51.5	55.1	68.1	70.9	99.6	95.6		
Goat farms	202.610	174.380	138.250	117.170	71.590	64.050	−68.4	−53.7
Goat flock	5.176.470	5.009.060	5.327.200	4.822.000	4.213.230	3.541.680	−31.6	−33.5
Goat/farm rate	25.5	28.7	38.5	41.2	58.9	55.3		
Grazed pasture (Ha)	657.940	583.850	605.280	824.250	2.450.240	1.859.250	182.6	207.2

Source: Our elaboration on EuroStat data

Table 6.4 Variations of sheep and goat farming data and rates in Spain (years 1990–2016)

Spain	1990	1995	2000	2005	2010	2016	% var. 1990–2016	% var. 2000–16
Farms with livestock			414.000	325.150	245.160	216.700		−47.7
Sheep farms	129.030	107.870	107.000	85.250	68.980	63.730	−55.7	−40.4
Sheep flock	17.500.850	19.019.300	20.926.770	19.660.060	16.574.220	15.862.160	−10.3	−24.2
Sheep/farm rate	135.6	176.3	195.6	230.6	240.3	248.9		
Goat farms	96.710	56.290	53.590	38.650	29.860	28.420	−70.6	−47.0
Goat flock	2.497.000	2.152.510	2.724.630	2.527.300	2.363.520	2.490.680	−0.3	−8.6
Goat/farm rate	25.8	38.2	50.8	65.4	79.2	87.6		
Grazed pasture (Ha)	8.448.400	8.199.100	9.368.390	8.653.210	8.377.390	7.615.990	−10.8	−18.7

Source: Our elaboration on EuroStat data

Table 6.5 Variations of sheep and goat farming data and rates in Italy (years 1990–2016)

Italy	1990	1995	2000	2005	2010	2016	% var. 1990–2016	% var. 2000–16
Farms with livestock				301.980	217.330	154.680		
Sheep farms	158.810	152.830	96.150	74.880	51.100	50.650	−74.9	−47.3
Sheep flock	8.721.620	10.667.970	6.808.330	6.991.140	6.782.180	7.026.540	−21.4	3.2
Sheep/farm rate	54.9	69.8	70.8	93.4	132.7	138.7		
Goat farms	87.330	75.190	48.070	30.960	22.760	21.710	−75.1	−54.8
Goat flock	1.246.520	1.372.940	922.660	917.850	861.940	982.000	−21.2	6.4
Goat/farm rate	14.3	18.3	19.2	29.6	37.9	45.2		
Grazed pasture (Ha)	4.106.080	3.758.220	3.418.080	3.346.950	3.434.070	3.233.230	−23.4	−5.4

Source: Our elaboration on EuroStat and Istat data

Table 6.6 Utilised agricultural area and permanent grassland in Italy, 2016

	Utilised agricultural area UAA (Ha)	Permanent grassland (Ha)	Permanent grass-land/ UAA (%)
Italy	1.143.960	319.690	27.9
North West	102.000	46.770	45.9
North East	177.730	49.850	28.0
Central Italy	178.460	52.000	29.1
Southern Italy	483.960	89.990	18.6
Islands	201.820	81.100	40.2

Source: Our elaboration on EuroStat data

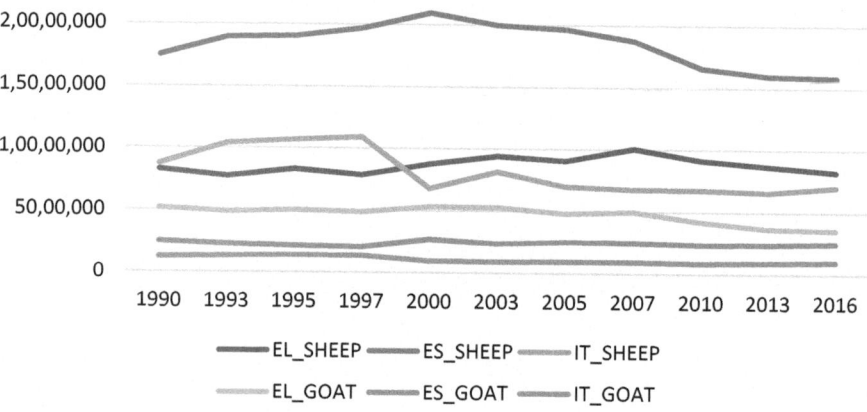

Fig. 6.6 Trends in national sheep and goat flocks in Greece, Spain and Italy (years 1990–2016). (Map legend: *EL* Greece, *ES* Spain, *IT* Italy. Source: Our elaboration on EuroStat data)

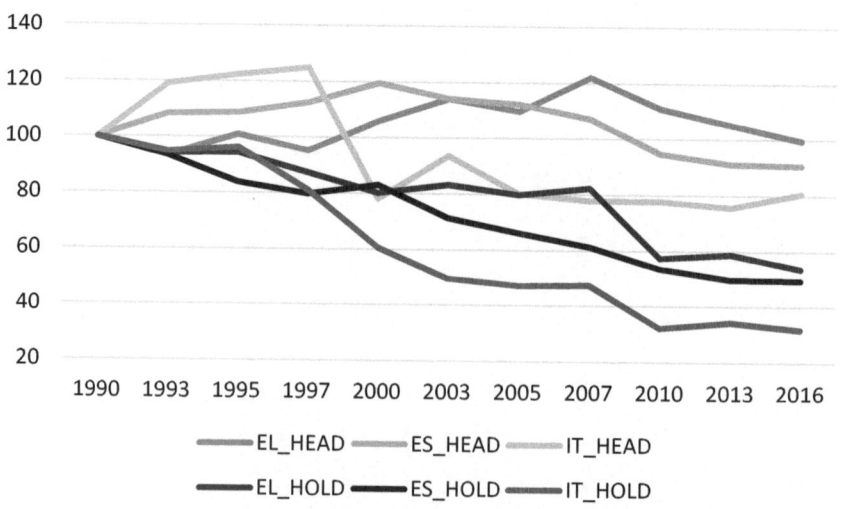

Fig. 6.7 Trends in national sheep flock and sheep farms in Greece, Spain and Italy (years 1990–2016 – index year: 1990 = 100). (Map legend: *EL* Greece, *ES* Spain, *IT* Italy, *HEAD* amount of sheep, *HOLD* amount of sheep farms. Source: Our elaboration on EuroStat data)

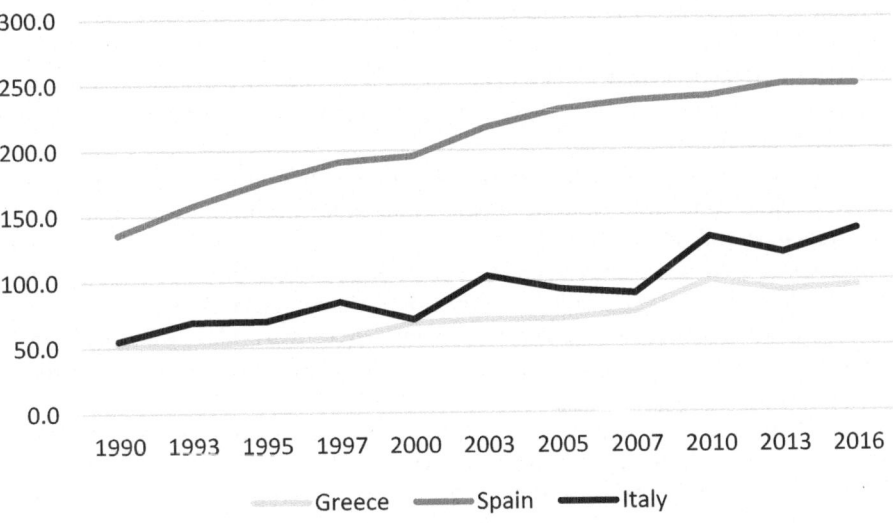

Fig. 6.8 Trend for average sheep farm size (average of sheep number for farm) in Greece, Spain and Italy (years 1990–2016). (Source: Our elaboration on EuroStat data)

We will use here some basic indicators referring to agro-pastoralism in Greece, Spain and Italy in recent decades, to assess the dynamics and trends that have characterised this sector in marginal rural areas and less industrialized regions of the recent decades in the EUMed. These data include the amount of agro-pastoral farms, the consistency of sheep and goats flocks (typically raised in extensive systems) as well as data related to the use of grazing lands.

In **Greece**, the livestock sector contributes to 28.3% of the total added value of national agriculture, and it has been particularly affected by recent declining trends. Small ruminant farming is the most important livestock production system in Greece; sheep and goats represent about 75% of the overall grazing units in the country, contributing significantly to local income and national GDP. Official data indicate that this sector accounts for 27% of the gross animal production value and 7.2% of the gross agricultural production value of the country (Hellenic Statistical Authority 2009).

It is specifically relevant in maintaining rural populations and as a main pattern of landscape management in the inner and island territories of the country, where employment opportunities outside livestock faming are limited. In these areas agro-pastoralism employs 17% of the workforce and accounts for 6.5% of the gross domestic product (Ragkos et al. 2016a, b).

According to Agricultural Census in Greece, livestock farms have decreased consistently in recent decades, with a drop of 39% from 2000 to 2016. Sheep are by far the most common livestock (representing about 38% of the total livestock population), but sheep farms have followed the same trends, although variations in sheep units have been changing through time in less linear ways Table 6.3). During the same period permanent grasslands increased consistently (+200%), indicating the growing relevance of extensive livestock breeding in some Greek regions during the recent decades.

Agro-pastoralism in Greece is territorially diversified: in Central Greece (Thessaly) as much as in the Northern, mountainous areas (Epirus) a mixed system of transhumance and semi-extensive sheep and goat breeding prevails. In lowlands more intensive sheep farming characterized by high investments and modern infrastructure has emerged especially since the 2000s. In most islands extensive grazing of mostly small ruminants represents a main source of livelihood for many rural communities, through the processing and the sale of Feta cheese through touristic networks (Ragkos et al. 2018).

In **Spain**, sheep and goat farming represent about a third of the overall livestock farm units in the country. Accession to EU in 1986 and corresponding financial flows had not interrupted the downward trend of traditional farming systems, particularly for livestock. The agricultural holdings with livestock decreased by 47,7% between 2000 and 2016 (Table 6.4). Data reveal that it was mostly smaller sized farms that disappeared from the landscape. Larger farms with more intensive livestock rearing have increased their stock of cattle and pig, while sheep and goat stock have decreased by 24,2% and 8,6% respectively. Territorial diversity is huge in Spain, from the northern Pyrenees, to central mountainous *mesta* systems, to drier pasturelands in Andalucia and Extremadura. Indications about change in pasturelands use are though similar throughout the country.

Satellite imagery confirms a −19% decline in grazed grasslands from 2000 to 2016 all over Spain, mostly due to changes in extensive livestock census, but also because of the different management patterns applied. A similar trend has been observed in dry mountainous areas, where the intensification of management in sheep farms is negatively related to (a) the use of natural grazing resources and to (b) the lack of generational renewal that risks compromising the continuity of most extensive farms. Depopulation and abandonment in these areas are reported to largely affect land use change, with significant impacts on environmental services and public goods (Caballero et al. 2009; Porqueddu et al. 2017).

In **Italy**, small ruminants breeding is a typical activity in most inner parts of the country, including in the Appenine and Alpine mountainous ranges, and in the islands. Within the period 1990–2016 agro-pastoral farms have undergone a process of concentration and modernization. Though the amount of sheep farms has dramatically dropped (−74,9%), during the same period the national sheep flock has decreased to a much lower degree (−21%). The average farm size (average of sheep number for farms) has shifted from almost 54,9 heads on average per flock to 138,7, almost three times more in less than two decades. In the same period, grazed pastures have undergone a gradual, continuous and consistent reduction (−23,4%, Table 6.5).

In 2016 grazing resources such as permanent grasslands and pastures still cover though large part of the agricultural land throughout the country − North West (32.5%), North East (24%), Southern Italy (20.6%) and Central Italy (18.5%), and about 40% of Sardinia and Sicily islands (Table 6.6). While agro-pastoral breeding of sheep, goats and even cattle are widespread in the large marginal territories of the country, territorial differences are many across regions and between Alpine, Apennine and insular settings.

6.4 Shepherding Labour and Immigrant Workforce in the Mediterranean Agro-pastoralism

Due to the extensive nature of agro-pastoralism, the work of the shepherd is intense and encompasses both physical labour as well as technical and managerial skills that range from knowledge about the climate, the vegetation, animal physiology and health, the ethology of predators, etc. (refer to Meuret 2010). Most of the shepherd's time is spent in harsh conditions, with limited access to public services, scarce connectivity and few opportunities for leisure and alternative activities. Continuous mobility and processing of milk add burdens to daily mansions, while the growing presence of predators and climatic vagaries represent further hardening factors. The shepherding profession relies thus on several technical and strategic skills, with specific ecological know-how as well as physical endurance.

In recent decades living conditions of shepherds have hardly improved, while working conditions have intensified, through a significant increase in their tasks and responsibilities. At the same time earnings have fallen as a result of competition on the international markets. Sectoral restructuring has thus contributed to creating unattractive conditions for the new generations, who have often decided not to follow their fathers' footsteps, and to avoid engaging in a profession with an uncertain perspective.

Through these lenses one can understand the crisis of agro-pastoral "vocation" and the relative lack of manpower and the problems of generational renewal on EU pasturelands in the Alps, Epirus, Apennines and Pyrenees, which rank among the areas most exposed to the risk of abandonment.

Figure 6.9 and Table 6.7 show that in Italy, Spain and Greece agro-pastoralist are younger than other farmers. Despite this, the problem of generational turnover appears evident, as over 50% of sheep and goats farmers are older than 55 years.

The challenge related to an aging population and limited generational renewal had already been identified as a priority by the Pastomed program, which in 2007

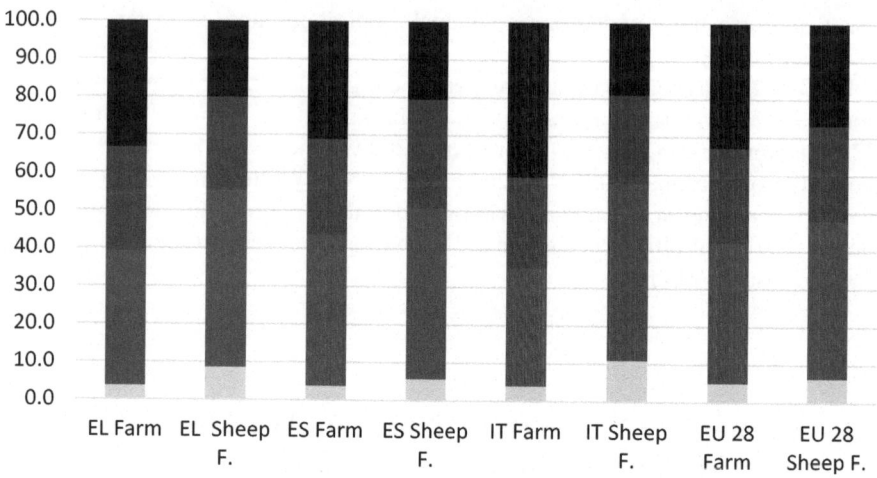

■ Less than 35 years   ■ From 35 to 54 years   ■ From 55 to 64 years   ■ 65 years or over

Fig. 6.9 Farm indicators by age of the manager in 2016 (comparison between *"average"* and *"sheep, goats and other grazing livestock"* farms). (Map legend: *EL Farm* Greece Farms, *EL Sheep F.* Greece "sheep, goats and other grazing livestock" farms, *ES Farm* Spanish Farms, *ES Sheep F.* Spanish "sheep, goats and other grazing livestock" farms, *IT Farm* Italian Farms, *IT Sheep F.* Italian "sheep, goats and other grazing livestock" farms. Source: Our elaboration on EuroStat data)

Table 6.7 Age of the farmers in 2016 and 2013

Country	Year	Less than 25 years	From 25 to 34 years	From 55 to 64 years	65 years or over
Average farmers					
Greece	2016	0.4	3.3	27.4	33.5
	2013	0.8	6.1	21.6	33.3
Spain	2016	0.2	3.6	25.4	31.2
	2013	0.4	4.9	25.6	29.7
Italy	2016	0.4	3.6	24	40.9
	2013	0.7	4.4	24.3	37.2
EU 28	2016	0.5	4.7	25	32.8
	2013	0.8	6.7	23.6	29.6
Sheep, goats and other grazing livestock farmers					
Greece	2016	0.7	7.9	24.6	20.2
	2013	2.7	14.5	19.6	18.1
Spain	2016	0.6	5.1	28.7	20.5
	2013	0.8	7.4	26.0	19.7
Italy	2016	1.2	9.8	23.1	19.1
	2013	2.2	11.5	21.4	16.9
EU 28	2016	0.7	5.7	25.0	26.9
	2013	1.1	8.0	23.9	24.4

Comparison between *"average"* and *"sheep, goats and other grazing livestock"* farms
Source: Our elaboration on EuroStat data

reported "the very high rate of over-55 compared with those under 35 years of age (. . .) and in many areas, the presence of elderly people 10 times more than young ones!" (Pastomed 2007:18). According to Slow Food, amongst local agricultural products in high danger of extinction, those from extensive livestock systems rank highest (dairy products are first, and meat second in many parts of southern Europe), due to the decrease of human resources (Essedra 2015).

As the local youth is decreasingly interested in working in this sector, the shepherding workforce has also changed, from family members to salaried labourers, and eventually from local workers to foreign ones. Today large parts of EUMed pasturelands are grazed by local flocks accompanied by immigrant shepherds who have come to fill this labour shortage, at a relatively low cost. Even the newly established agro-pastoral farms and enterprises often rely on immigrants' workforce (Circerchia and Pallara 2009; Cicerchia 2012, 2014; Nori and de Marchi 2015).

In Greece there has been a massive influx of immigrants to rural areas in recent decades, especially following the collapse of the Albanian regime and the consequential borders opening in the early 1990s. Important proportions of Albanians came to live and work in mountainous villages of northern Epirus, contributing significantly to the restructuring of the extensive livestock sector and to the local social, economic and demographic fibres (Kasimis and Papadopoulos 2005; Kasimis 2010). These early flows slowly opened the way to shepherds originating from Eastern Europe (Bulgaria), and more recently to workers from Eastern Asia (India and Pakistan). In the Epirus and Peloponnesus immigrants represent today about half of the agro-pastoral salaried workforce; in Crete they account for about a third (Ragkos et al. 2013; Nori 2018).

By providing a skilled workforce at a relatively low cost, immigrant shepherds have contributed to ensuring the maintenance and the reproduction of agro-pastoral enterprises. Due to the shortage of family labour, the recruitment of immigrants has enabled farm women to maintain their domestic role, and younger household members to continue studying and/or looking for employment opportunities outside the agricultural sector (Papadopoulos and Roumpakis 2009; Ragkos et al. 2016a). These contributions have supported on one side the development of large, innovative and specialized dairy farms along the schemes proposed by the CAP, while on the other side they have also enabled the maintenance of more traditional transhumance systems that characterize agro-pastoral resource management in certain parts of the country.

In Spain, immigration from a variety of countries has contributed to the labour reconfiguration of several local agro-pastoral systems. In areas where predation is encroaching, the presence of shepherds is becoming increasingly important to take care of flocks. For example, in the north-eastern Pyrenees, immigrants constitute about half of the salaried shepherding workforce. Traditionally these immigrants originate from Morocco and Romania, but more recent trends indicate a growing presence of shepherds issued from Bulgaria, Ukraine, and increasingly Sub-Saharan and Latin-American workers. The ratio of immigrant shepherding workforce decreases to one in three workers in central Spain, in the Castillas, as well as in Galicia and Extremadura, where Portuguese workers are more likely to be found (Nori 2017a, b). Some of these workers have joined some form of training in one of the six pastoral regional schools present in the country.

In Italy the salaried workforce in livestock production is largely and increasingly composed of immigrants, due to the difficulty of recruiting local people. While the presence of immigrant workers is well reported in more intensive livestock farming (Lum 2011), their contribution to the agro-pastoral systems have been only limitedly appreciated (Nori 2016; Farinella et al. 2017).

Immigrants constitute today about two-thirds of the agro-pastoral salaried work-force in most Alpine and Apennine pastures in Italy, where the growing presence of predators has contributed to the reincorporation of shepherds in many inner areas of the country (Nori and de Marchi 2015). In Abruzzo, a region with an important pastoral tradition, official data indicates that nine over ten salaried shepherds are Macedonians or Romanians (Coldiretti 2010). In Sardinia, where agro-pastoral land use covers most of the island territory and whose small ruminants' stock represents over 40% of the national flock, one every three salaried shepherds is a foreigner, representing a critical resource for reproducing the family farming enterprise (Farinella and Mannia 2017, 2018; Nori 2018; Farinella 2019b).

Box: Shifting Workforce – The Emblematic Example of Sardinian Pastoralism

In the past being a salaried shepherd (*su theracu* in Sardinian language) represented a common step in the socio-economic career of local youth, before generating own capacities and money to raise a herd on your own right (Farinella and Mannia 2017, 2018). Today conditions have made this job not attractive for young people who rather prefer emigrating elsewhere in search of employment. It is today up to immigrant workforce to perform the functions related to livestock management and breeding, but also collateral tasks such as clearing lands, building fences, collecting timber, farming animal feed, producing cheese, as well as building or mechanical activities on the farm. By triangulating data from different sources on resident and working populations, estimates attest to about a thousand the Romanians employed in agro-pastoralism in the region in 2016, mostly engaged in lowland medium-sized, semi-intensive sheep farms (Farinella and Mannia 2017, 2018). In the last two years the number of Romanians have decreased, during the milk crisis (Farinella 2019a, b; Simula 2019).

Most immigrate shepherds originate from other pastoral communities in the Mediterranean, and thus show some skills and capacities related to livestock husbandry. In some regions a process of substitution of a group over another is reported, whereby the Albanians, Macedonians and Moroccans that started working as shepherds in the 1990s have been replaced by Romanians in the 2000s, due to the latters' accession to the EU, and the related facilitation in administrative and mobility terms. More recently an increase of workforce originating from sub-Saharan Africa as well as Eastern Asia is being reported (Nori and López-i-Gelats 2017).

Data and trends in Portugal and France are similar, although the dynamics of immigrants' shepherding workforce have followed different patterns and trajectories. Other agricultural sub-sectors specific to marginal territories show similar dynamics; for example, workers from Eastern Europe and the Balkans account for about 40% of the forestry workforce in central Italy, and in many cases they provide a crucial contribution to maintain longstanding local, traditional sylvicultural systems (Cicerchia and Pallara 2009; Luisi and Nori 2017).

6.5   The Conditions of Immigrant Shepherds

The typical profile of migrants that have come to work as a shepherd in EUMed region is that of a male, aged between 25 and 40, native of a country of the Mediterranean (predominantly from Romania, Morocco, Albania or northern Macedonia), but recently also from Asia (e.g. Pakistan, India), sub-Saharan Africa (e.g. Ghana, Senegal) and even from Latin America (in Spain). Though not necessarily from pastoral areas, the large majority comes from a rural setting and has direct experience in livestock breeding:

> We are organised and upon demand we can seek for more workers from our networks, mostly in our villages in north-eastern Romania. There, everybody used to keep sheep. Most households produced their own cheese, that is where we have learnt. We know how to deal with sheep. (TRAMed Interview with a shepherd, Triveneto, April 2015)

History, language, and the networks of migrants have shaped the different migratory patterns. Romanians are mostly found in Italy and parts of Spain, Moroccans in parts of Spain and southern France, and Albanians in Greece. Socio-cultural differences aside (e.g. Orthodox or Muslim in predominately Catholic societies), immigrant shepherds are generally appreciated for their technical skills, as well as for their endurance, flexibility and adaptability, in that they accept the working conditions and salary generally rejected by the local population; "they are like us 60 years ago" (Nori and López-i-Gelats 2017). Younger shepherds are preferred as they are more likely to learn local languages and follow indications.

Average immigrant shepherds work individually and live in isolated sheepfolds, often in remote areas far from villages and with limited means to move. Cases exist in certain areas where shepherds are seasonal workers, who tend to return home or to work elsewhere when the peak season is over (ie. once the transhumance or the intense milking periods are over).

Salary rates normally range between 600 and 1000 euros per month, for a full-time engagement, with very limited free time and little holiday. In addition to the salary bed and board are often provided by the farm, though often associated to the sheepfolds. This arrangement enables farmers to underpay workers and to maintain forms of control on them (Farinella and Mannia 2018). Immigrants' revenue is often invested in their home communities, at times on the purchase of family land and livestock, with the hope they would one day get back; this also results from having limited chances to graduate and remain in destination areas.

The contractual arrangements are often quite informal, partial and precarious. Conditions of illegality, limited rights, scarce salary and poor living and working standards represent typical features of workers operating in this grey context, on the margins of a rural world that is already marginal on its own. There are no trade unions, recruitment is carried out exclusively by word of mouth through personal networks and individual arrangements that presents at times exploitative mechanisms (Farinella 2019b).

These elements add to a situation where limited access to land and credit are the main factors inhibiting the capacity and the interest of these workers to remain in this sector. This is further exacerbated by constraints related to residence permits, entrepreneurial licenses and overall citizenship rights, including compliance with CAP procedures and rules, which would enable them accessing precious financial support. In this context, workers see little chance for improving their socio-economic conditions. They often remain a few months or years in this sector, switching between different farms in search of more comfortable living and working conditions. However, the incentives to take over existing farms or establishing new ones remain limited.

The limited formalization of contractual relationships, poor labour conditions and the very scanty prospects for a socio-economic "upgrade" are complementary and inter-twined elements that characterize most Euro-Mediterranean agriculture (Pittau and Ricci 2015; Farinella and Mannia 2018).

The fact that a generational change is accompanied by an ethnic one is not new to the region. Over the last century, Mediterranean pastoralism has witnessed Sardinians colonising abandoned pasturelands in central Italy, southern Spanish herders moving to graze the Pyrenees, northern Italian shepherds migrating to Provence and Switzerland, the moves of Valachos and Arvanites flocks and shepherds throughout Greece and Kurdish shepherds in several regions of western Turkey (Lebaudy 2010; Meloni 2011; Nori 2016 – cfr. Table 6.8 and Fig. 6.10). These communities have substantially contributed to keeping pasturelands of destination countries populated, alive and productive.

Compared to past migratory dynamics, today an important difference relates to the difficulties faced by immigrants to graduate socially and economically in their activity. Migrants that have recently come to operate as shepherds in the EUMed have barely improved through time their conditions as workers and much as citizens. Their transition from workers to entrepreneurship and livestock farmers in their own

Table 6.8 Migratory flows of shepherds through the Mediterranean in the twentieth century

Destination region	Origin of migrant shepherds				
	Late 1800	1950s	1980s	1990s	2000s
Provence	Italy and Spain	Morocco and Tunisia			Romania
Central Italy		Sardinia	Morocco and Tunisia	Albania, FYROM	Romania
Pyrenees	Neighbouring valleys	Andalusia		Morocco	Romania, sub-Saharan Africa
Turkey		Kurdistan		Afghanistan	

Source: TRAMED data elaboration

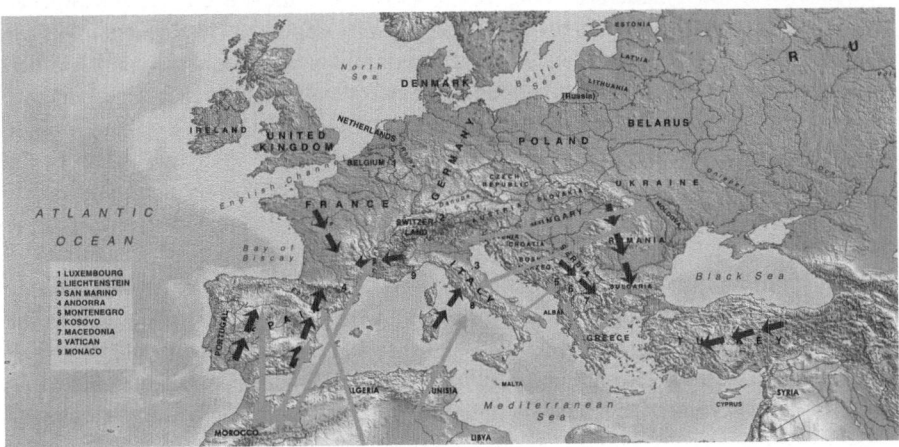

Fig. 6.10 Trajectories of past (darker) and present (lighter) patterns of shepherds' migrations in the Mediterranean. (Source: TRAMED data elaboration)

right shows very low rates. This in turn constrains the capacity of the incoming population to sustainably contribute to repopulate pastures and to reproduce agro-pastoralism in the longer term.

6.6   Stories of Immigrants in Italian Agro-pastoralism

In this section, some personal stories from the Italian agro-pastoralism are detailed, so to provide a human perspective to the dynamics and processes assessed also with a view to focus on people, individual stories and personal projects.

Italy is a country where *Made in Italy* agro-food represents a strategic sector and plays a critical economic role through agricultural export and the tourism industry. The Italian dairy sector is representative of the contribution of immigrants to the globally recognised excellence of this sector.[1] The agro-pastoral immigrant work-force is in fact not only relevant in terms of taking care of herds and flocks, but as well in the dairy processing industry. Apart from the better-known case of the Sikh community with the Parmesan (Lum 2011; Azzeruoli 2014), and the Bengali communities for the buffalo mozzarella, foreign communities play a strategic contribution in the value chains of Fontina and Pecorino cheeses, which are typical regional products issued from pastoral settings. Most representative "made in Italy" cheeses are in fact made by immigrants' hands.

In Valle d'Aosta, almost two thirds of the workers employed in cattle breeding are foreigners. From taking care of the local breed cattle to the processing of milk in the alpine huts (*malghe*), immigrants largely contribute to the production of the famous Fontina cheese that characterises the region. Formerly almost exclusively Moroccan shepherds, in recent years they have been partially replaced by Romanians. Data from 2014 reported 303 non-EU workers (predominantly Moroccans) and 335 foreign EU workers (predominantly Romanians) officially employed, together with several irregular workers (around 100). Foreign labourers have more than doubled over the last two decades, representing to date about more than two thirds of the salaried shepherding workforce. Living and working conditions are quite harsh, these being main reasons local inhabitants do not seem interested in undertaking this work. In such terms immigrant shepherds do not compete thus with local workers, and their contribution can therefore be said to be essential for maintaining the traditional system of breeding for the production of Fontina DOP (Cicerchia 2014; Nori 2016, 2018).

In Sardinia, usually, Romanians are involved only in the sheep management and in the milking activities, as most shepherds produce milk for processing industries.

---

[1] http://slowfood.com/resistenzacasearia/ita/77/vullnet-alushani-caciocavallo-podolico-del-gargano-italia

They often work in pairs and are regulated with seasonal contracts, which expire in summer, when sheep milking is over. Some of them engage in complementary activities to supplement their salary, such as vineyard pruning, harvesting wood and collecting cork.

In multifunctional farms, where sheep-breeding is associated to other activities (agritourism, educational farms, other breeding and agricultural activities) and farmers produce artisanal cheeses in local dairies, Romanians are also directly involved in the dairy processing, including the traditional Fiore Sardo and Pecorino Sardo (Farinella 2019b).

In Sardinia and in other Italian regions, exceptional cases exist where foreign shepherds have engaged and succeeded in *scaling up to livestock ownership and farm management*. In these experience immigrant shepherds look into opportunities to set up their own flocks, and/or cooperate amongst themselves or with local dwellers in sharing land, subsidies or credit assets (Farinella and Mannia 2017; Farinella et al. 2017; Nori 2018).

For instance, M., a Romanian shepherd who came to Italy 10 years ago, initially worked without a contract or insurance. Seven years ago, he got a contract which finally made him eligible for Italian citizenship, which is needed to register as an entrepreneur and to legally own a flock. With his savings, he was able to accumulate a few animals each year, which he kept within the flock of his employer. Recently, he and his employer have been talking about jointly managing a common flock. They plan to share the costs and responsibilities, as well as the profits. With an established business, M. will be able to bring his wife and children to Italy. Other examples of such socio-economic graduation exist where two immigrants have shared resources and responsibilities or in areas where pasture lands are communal and therefore more easily accessible.

Another similar case is the story of G., a Romanian shepherd in Sardinia. G. started as a worker in a sheep farm. After a short time he decided to try to start his own business using the networks that he created in the area and the relationship of trust with his employer, with whom they started a cooperation. Following local practices whereby farms could be split amongst siblings or relatives to enhance and extend CAP support. G. started his own activity, though he still works with his previous Sardinian employer. The Romanian shepherd continues to live in a sheepfold house, made available by his now Sardinian colleague. All the activities (bringing the animals to the pasture, the milking and the cultivation of the land) continues to be carried out together by the Sardinian and the Romanian shepherd, using the structures of the former. However, G. is no longer an employee, he does not receive a salary; he is in charge of his own milk, while also receiving CAP support. He does not pay the rental, but he is charged with his work for the land and structures he utilizes. This is an example of very interesting cooperation, which is based on local forms of reciprocity and non-monetised exchanges. In this way positive externalities are created for both individuals as well as for the local community they live in. G. has the dream to save to be able to buy a piece of land in Sardinia.

Another example of multi-functional rural entrepreneurship by migrants is that of the organic farm *"La Capra Felice"*, launched in Trentino in 2010 by a young Ethiopian woman. After studying in Italy, she returned to her country of origin to create a project of sustainable agriculture, but she had eventually been forced to leave the country because of its struggle against the excessive power of food multinationals. When she returned to Italy, she started a multifunctional agricultural activity, in which she raises goats of the native breed, the Mòchena spotted, at risk of extinction and other locally typical animals.

She produces directly and sells locally her own diversified homemade organic sheep-goat cheeses, through short supply chain networks. Some of her dairy products have gained the Slow Food dairy resistance award. The choice of an extensive breeding, based on wild grazing and on particularly frugal native breeds, allows the milk to have a high organoleptic quality (which derives from the biodiversity of natural pastures) and to organize a diversified agricultural activity, which sustainable in either economic terms as well as agro-ecological terms. She organizes training activities and educational farming with the primary school and contribute to the production of services and welfare in rural areas, for example through the reception of a refugee (Sivini 2019).

6.7   Conclusions

Agro-pastoralism represents an increasingly appreciated practice in Europe and elsewhere for its quality products as well as for the important public goods it contributes, in the form of socio-ecosystem services. These recognitions are rewarded through public subsidies and market pricing. Despite growing societal acknowledgement and appreciation, agro-pastoralism is though decreasingly practiced by the local populations; the consistent decrease in farms and flocks holds important consequences on the livelihood systems as well as on the natural resource management of marginal territories.

Immigrants have proved to be a strategic resource to fill the shepherding labour gaps left by the declining, ageing and decreasingly interested local population. Notwithstanding the important contributions received through public policies, and no matter what type of entrepreneurial strategy pursued to cope with the sector restructuring, immigrant shepherds have shown to represents a strategic asset for this agricultural practice, by providing a skilled labor force at a relatively low cost. Without foreign workers, many agro-pastoral farms would face today great difficulty in pursuing their activities and in so maintaining EUMed marginal territories alive and productive.

Yet oftentimes the conditions under which these dynamics are taking place are not enabling the sustainable reproduction and development of agro-pastoralism in the region, as the conditions and the opportunities immigrant shepherds enjoy are not conducive towards their integration into agro-pastoralism in more consistent and structured ways. In a lose-lose situation, immigrant shepherds do not graduate in socio-economic terms thus to evolve into farmers themselves, while elderly farmers do not find people capable of taking over their farms when they retire. Society at large witnesses the disappearance of flocks, the abandonment of marginal lands, together with the loss of quality products and services, despite important degrees of public investments.

This case reflects more in general the complexity and contradictions of immigrant presence in rural regions and in the agriculture sector – in that foreign workers are willing to accept working conditions and salaries usually rejected by the local people, their presence is only temporary, with relevant implications for local patterns of rural development.

The cases where shepherds have graduated to livestock farmers represent rare exceptions, and important opportunities to capitalise upon, with a view to secure a more sustainable rural society.

Appendix: The Presence of Immigrants in EUMed Agro-pastoralism

Region	Main production	% foreign on total salaried shepherd	Origin country of most of them	Average monthly salary (€)	Source
Italy					
Abruzzo	Milk	90%	Macedonia, Romania, Albania	800	Coldiretti (2010)
Triveneto	Meat	70%	Romania	800	TRAMed
Piedmont	Meat and milk	70%	Romania, Moldavia	800	TRAMed; Cicerchia and Pallara (2009) and Cicerchia (2014)
Val d'Aosta	Milky cows	70%	Romania, Morocco	2000	Cicerchia (2014)
Sardinia	Milk	35%	Romania, Morocco	500–600	Farinella and Mannia (2017); TRAMed
Calabria	Milk	35%	Kurdistan, Pakistan, India	500–600	Cicerchia and Pallara (2009)
Greece					
Thessaly	Milk	50%	Albania, Bulgaria, Romanian Vlachs	400–600	Thales, Domestic

(continued)

Region	Main production	% foreign on total salaried shepherd	Origin country of most of them	Average monthly salary (€)	Source
Peloponnese	Milk	40%	Albania, Bulgaria, India, Pakistan	400–600	Thales, Domestic
Crete	Milk	35%	Albania, Bulgaria, India, Pakistan	400–600	Thales, TRAMed
France					
Provence	Meat	Mostly during winter for large flocks	Romania Morocco, Tunisia	1400	TRAMed; Fossati (2015)
		Mostly on summer pastures	Other regions of France or Northern Europe	1500–2500	TRAMed; Meuret (2010)
Pyrenees	Milk	Few salaried shepherds	Quite limited phenomenon		Meuret (2010)
Maritime Alps	Meat	20%	Romania		TRAMed
Corse	Milk and meat		Morocco		Terrazzoni (2010)
Spain					
Valencia Community		70%	Morocco	600	AVA (2009)
Catalan Pyrenees	Meat	55%	Romania, sub-Saharan Africa	6–700	Nadal et al. (2010)
Aragon Pyrenees	Meat	60%	Morocco, Romania, Bulgaria, Ukraine		TRAMed;
Andalucia			Romania, sub-Saharan Africa		TRAMed;
Castillas	C. Léon meat C. Mancha milk	35%	Morocco, Romania, Bulgaria, Portugal		TRAMed; Plataforma
Basque country	Milk		Romania	1000	TRAMed;
Galicia			Portugal		TRAMed;
Extremadura			Portugal		TRAMed;

Source: TRAMed project

References

Agnoletti, M. (Ed.). (2013). *Italian historical rural landscapes. Cultural values for the environment and rural development.* Dordrecht: Springer.

Antrop, M. (1997). The concept of traditional landscapes as a base for landscape evaluation and planning. The example of Flanders. *Landscape and Urban Planning, 38*(1–2), 105–117. https://doi.org/10.1016/S0169-2046(97)00027-3.

Antrop, M. (2005). Why landscapes of the past are important for the future. *Landscape and Urban Planning, 70*(1–2), 21–34. https://doi.org/10.1016/j.landurbplan.2003.10.002.

A.V.A. (2009). Una Jornada sin Inmigrantes. Informe. Asociación Valenciana de Agricultores http://www.levante-emv.com/sociedad/2009/12/06/jornada-inmigrantes/658140.html

Azzeruoli, V. (2014). *Legami tra pianure. Gli intermediari nella migrazione dei panjabi indiani in Italia.* Phd thesis, Univeristy o Padua. http://paduaresearch.cab.unipd.it/7004/

Beaufoy, G., & Ruiz-Mirazo, J. (2013). Ingredientes para una nueva Política Agraria Común en apoyo de los sistemas ganaderos sostenibles ligados al territorio. *Revista Pastos, 43*(2), 25–34.

Braudel, F. (1982, V ed., ed.or.1949). *La Méditerranée et le monde méditerranéen à l'époque de Philippe II.* Paris: Armand Colin.

Brisebarre, A. M. (2007). *Bergers et transhumances.* Romagnat: De Borée.

Caballero, R., Fernández-González, F., PérezBadia, R., Molle, G., Roggero, P. P., Bagella, S., D'Ottavio, P., Papanastasis, V. P., Fotiadis, G., Anna, S. A., & Ioannis, I. I. (2009). Grazing systems and biodiversity in Mediterranean areas: Spain, Italy and Greece. *Revista Pastos, 39*(1), 9–152.

Campbell, J. K. (1964). Honour, family and patronage. In *A study of institutions and moral values in a Greek mountain community.* Oxford: Clarendon.

Cicerchia, P. (Ed.). (2012). *Indagine sull'impiego degli immigrati in Agricoltura in Italia.* Roma: INEA.

Cicerchia, P. (Ed.). (2014). *Indagine sull'impiego degli immigrati in Agricoltura in Italia.* Roma: INEA.

Cicerchia, M., & Pallara, P. (Eds.). (2009). *Gli immigrati nell'agricoltura italiana.* Roma: INEA. http://www.red-network.eu/resources/toolip/doc/2011/11/04/inea%2D%2D-gli-immigrati-nellagricoltura-italiana%2D%2D-2009.pdf

Coldiretti. (2010). *Complemento al Dossier immigrazione Istat 2010.* https://www.coldiretti.it/archivio/istat-nelle-stalle-dove-si-ottiene-il-latte-per-il-parmigiano-reggiano-1-lavoratore-su-3-e-indiano-12-10-2010

ELSTAT. (2009). *Hellenic statistical office.* Athens: Coincise Statistical Yearbook.

Essedra. (2015). *Environmentally sustainable socio-economic development of rural areas.* Slow Food project funded by the EU Instrument for Pre-accession Assistance (IPA) Civil Society Facility.

Eurostat. (2016). *Agriculture, rural development statistics.* Luxembourg: Eurostat.

Farinella, D. (2018). La pastorizia sarda di fronte al mercato globale. Ristrutturazione della filiera lattiero-casearia e strategie di ancoraggio al locale. *Meridiana, 93.* https://www.viella.it/rivista/9788833131597/4251

Farinella, D. (2019a). The case of the Pecorino Romano dairy production chain in Sardinia, Italy. In M. Migliorini (Ed.), *FOOD TRACK. A transparent and traceable supply chain for the benefit of workers, businesses and consumers: The role of a multisectoral approach to industrial relations and corporate social responsibility* (pp. 130–163). Rome: CGIL.

Farinella, D. (2019b). Le ragioni e la lotta dei pastori sardi. *Gli Asini, 62,* 17–22.

Farinella, D., & Mannia, S. (2017). Migranti e pastoralismo. Il caso dei servi pastori romeni nelle campagne sarde. *Meridiana, 88,* 175–196. https://www.viella.it/rivista/9788867288601/3970

Farinella, D., & Mannia, S. (2018). «Mi chiamo Serban e non sono il romeno di nessuno, sono il romeno di me stesso». Pratiche di assoggettamento e soggettivazione tra pastori sardi e servi pastori romeni. *Etnografia e ricerca qualitativa, 3,* 405–426. https://doi.org/10.3240/92124.

Farinella, D., & Meloni, B. (2013). Dalla tradizione all'innovazione: prospettive e opportunità delle filiere agroalimentari territorializzate. In B. Meloni & D. Farinella (Eds.), *Sviluppo rurale alla prova, dal territorio alle politiche* (pp. 127–154). Torino: Rosenberg & Sellier.

Farinella, D., Nori, M., & Ragkos, A. (2017). Change in Euro-Mediterranean pastoralism: Which opportunities for rural development and generational renewal? In C. Porqueddu, A. Franca, G. Molle, G. Peratoner, & A. Hokings (Eds.), *Grassland reources for extensive farming systems in marginal lands: Major drivers and future scenarios* (Volume 22 grassland science in Europe) (pp. 23–36). Wageningen: Wageningen Academic Publishers.

Fossati, L. (2015). De l'émigration à l'immigration. Savoir-faire berger en Valle Stura di Demonte. In: G. Lebaudy, B. Msika, B. Caraguel (editors), L'alpage au pluriel, Cardère éditeur.

Fréve, E. R. (2014). L'élevage ovin français : entre finalité domestique et mission de service public, la transformation du métier de berger en Provence. In G. Gallenga (Ed.), *De la porosité des secteurs public et privé, Une anthropologie du service public en Méditerranée* (pp. 33–50). Aix-en-Provence: Presse Universitaire de Provence.

Hadjigeorgiou, I. (2011). Past, present and future of pastoralism in Greece. *Pastoralism: Research, Policy and Practice, 1*, 24.

IFAD. (2017). *Sending money home: Contributing to the SDGs, one family at a time.* Rome: International Fund for Agricultural Development (IFAD).

ISMEA. (2017). *Settore ovicaprino, Scheda di settore.* http://www.ismeamercati.it/lattiero-caseari/latte-derivati-ovicaprini

Kasimis, C. (2010). Demographic trends in rural Europe and migration to rural areas. *Agri Regioni Europa, 6/21.* https://agriregionieuropa.univpm.it/it/content/article/31/21/demographic-trends-rural-europe-and-international-migration-rural-areas

Kasimis, C., & Papadopoulos, A. G. (2005). The multifunctional role of migrants in the Greek countryside: Implications for the rural economy and society. *Journal of Ethnic and Migration Studies, 31*(1), 99–127. https://doi.org/10.1080/1369183042000305708.

Kerven, C., & Behnke, R. (2011). Policies and practices of pastoralism in Europe. *Pastoralism: Research, Policy and Practice, 1*, 28. https://doi.org/10.1186/2041-7136-1-28.

Le Lannou, M. (1979). *Pastori e contadini di Sardegna.* Cagliari: Edizioni della Torre.

Lebaudy, G. (2010). Shepherds from Piedmont in Provence: Career paths and mobility. In *Proceedings of the XVI international oral history conference between past and future: Oral history, Memory and meaning.* Prague.

López-i-Gelats, F. (2013). Is mountain farming no longer viable? The complex dynamics of farming abandonment in the pyrenees. In S. Mann (Ed.), *The future of mountain agriculture* (Springer geography). Berlin: Springer. https://doi.org/10.1007/978-3-642-33584-6_7.

Luisi, D., & Nori, M. (2017). Gli immigrati nella Strategia Nazionale per le Aree Interne, dalle Alpi agli Appennini. In A. Membretti, I. Kofler, & P. P. Viazzo (Eds.), *Per forza o per scelta. L'immigrazione straniera nelle Alpi e negli Appennini.* Roma: Aracne.

Lum, K. D. (2011). *The quiet Indian revolution in Italy's dairy industry.* Firenze: European University Institute.

Marsden, T. (1995). Beyond agriculture? Regulating the new rural spaces. *Journal of Rural Studies, 11*(3), 285–296. https://doi.org/10.1016/0743-0167(95)00027-K.

Mattalia, G., Volpato, G., Corvo, P., & Pieroni, A. (2018). Interstitial but resilient: Nomadic Shepherds in Piedmont (Northwest Italy) amidst spatial and social marginalization. *Human Ecology, 46*, 747–757. https://doi.org/10.1007/s10745-018-0024-9.

Mattone, A., & Simbula, P. (Eds.). (2011). *La pastorizia mediterranea Storia e diritto (secoli XI-XX).* Roma: Carocci.

McNally, S. (2001). Farm diversification in England and Wales. What can we learn from the farm business survey? *Journal of Rural Studies, 17*(2), 247–257. https://doi.org/10.1016/S0743-0167(00)00050-4.

Meloni, B. (1984). *Famiglie di pastori: continuita e mutamenti in una comunita della Sardegna Centrale 1950–1970.* Torino: Rosenberg & Sellier.

Meloni, B. (2011). Le nuove frontiere della transumanza e le trasformazioni del pastoralismo. In A. Mattone & P. F. Simbula (Eds.), *La Pastorizia Mediterranea* (pp. 1051–1076). Roma: Carocci.

Meloni, B., & Farinella, D. (2015a). L'evoluzione dei modelli agropastorali in Sardegna dagli anni cinquanta ad oggi. In L. Marrocu, F. Bachis, & V. Deplano (Eds.), *La Sardegna contemporanea* (pp. 447–473). Roma: Donzelli.

Meloni, B., & Farinella, D. (2015b). Pastoralismo e filiera lattiero casearia, tra continuità ed innovazione: un'analisi di caso. *Meridian, 84*, 1–26.

Meuret, M. (2010). *Un savoir-faire de bergers.* Versailles: Editions Qua Beaux livres.

Moreira, O., Carolino, N., & Beloet, C. (2016). Climatic changes: Scenarios and strategies for the livestock sector in Portugal. In CIHEAM. *Watch Letter 37 – Mediterranean agriculture and climate change. Impacts, adaptations, solutions.* https://www.ciheam.org/uploads/attachments/278/WL_37_PDF_Complet.pdf

Nadal, S. E., Ricou, I. J., & Estrada, B. F. (2010). Transhumàncies del segle XXI. La ramaderia ovina i la transhumància a l'Alta Ribagorça. *Temes d'Etnologia de Catalunya, 20.*

Nori, M. (2015). Pastori a colori. *Agri Regioni Europa 11/43.* https://agriregionieuropa.univpm.it/it/content/article/31/43/pastori-colori

Nori, M. (2016). Shifting Transhumances: Migrations patterns in Mediterranean pastoralism. In CIHEAM. *Watch Letter 36 – Crise et resilience en la Mediterranee.* www.iamb.it/share/integra_"les_lib/"les/WL36.pdf

Nori, M. (2017a). Migrant Shepherds: Opportunities and challenges for Mediterranean. *Pastoralism Journal of Alpine Research, 105/4.* https://rga.revues.org/3544

Nori, M. (2017b). *Immigrant Shepherds in Southern Europe.* E-paper, Heinrich Böll Stiftung Foundation. https://www.boell.de/en/agriculture-food-production-and-labour-migration-south ern-europe

Nori, M. (2018). *Agriculture and rural territories in the Mediterranean: The case for mountainous communities.* In *MEDITERRA 2018 – Inclusion and migration challenges around the Mediterranean.* Paris: CIHEAM.

Nori, M., & de Marchi, V. (2015). Pastorizia, biodiversita e la sfida dell'immigrazione: il caso del Triveneto. *Culture della sostenibilità, 15,* 78–101.

Nori, S., & Gemini, M. (2011). The common agricultural policy vis-à-vis European pastoralists: Principles and practices. *Pastoralism, 1,* 27. https://doi.org/10.1186/2041-7136-1-27.

Nori, M., & López-i-Gelats, F. (2017). *Relevo generacional e inmigrantes en el mundo pastoril: el caso del Pirineo catalán.* Paper presented at the CSIC conference in Madrid.

Nori, M., Ragkos, A., & Farinella, D. (2017). Agro-pastoralism as an asset for sustainable Mediterranean Islands. In K. Jurcevic, L. Kaliterna Lipovcan, & O. Ramljak (Eds.), *Mediterranean Issues, Book 1, Imagining the Mediterranean: Challenges and perspectives* (pp. 135–147). Vern: Institute of Social Sciences Ivo Pilar.

Papadopoulos, T., & Roumpakis, A. (2009). *Familistic welfare capitalism in crisis: The case of Greece* (ERI Working Paper Series, WP-09-14). ERI, University of Bath, UK.

Pastomed. (2007). *Le pastoralisme méditerranéen, situation actuelle et perspectives: modernité du pastoralisme méditerranéen.* Rapport final du projet Interreg III PastoMED, Manosque.

Pernet, F., & Lenclud, G. (1977). *Berger en corse: essai sur la question pastorale.* Grenoble: PUG.

Pittau, F., & Ricci, A. (2015). Agricoltura e migrazione nel contesto dei muovi mercati globali. *Dialoghi mediterranei, 12.* www.istitutoeuroarabo.it/DM/agricoltura-e-immigrazione-nel-contesto-dei-nuovi-mercati-globali/

Pitzalis, M., & Zerilli, F. (2013). Il giardiniere inconsapevole. Pastori sardi, retoriche ambientaliste e strategie di riconversione. *Culture della sostenibilità, 6,* 12.

Porqueddu, C., Franca, A., Molle, G., Peratoner, G., & Hokings, A. (2017). *Grassland reources for extensive farming systems in marginal lands: Major drivers and future scenarios* (Volume 22 grassland science in Europe). Wageningen: Wageningen Academic Publishers.

Ragkos, A., & Nori, M. (2016). The multifunctional pastoral systems in the Mediterranean EU and impact on the workforce. *Options Méditerranéennes, Série A. Séminaires Méditerranéens* 114 (15). Proceedings of the FAO-CIHEAM workshop Ecosystem services and socio-economic benefits of Mediterranean grasslands, Orestiada, Greece.

Ragkos, A., Mitsopoulos, I., Siasiou, A., Skapetas, V., Kiritsi, S., Bambidis, V., Lagka, V., & Abas, Z. (2013). Current trends in the transhumant cattle sector in Greece. *Scientific Papers Animal Science and Biotechnologies, 46*(1), 422–426.

Ragkos, A., Siasiou, A., Galanopoulos, K., & Lagka, V. (2014). Mountainous grasslands sustaining traditional livestock systems: The economic performance of sheep and goat transhumance in Greece. *Options Méditerranéennes, 109*, 575–579.

Ragkos, A., Koutsou, S., Tsivara, T., & Manousidis T. (2016a). The operation of Pomak livestock farms in Northern Evros, Greece. *Options Mediterraneennes, Serie A: Mediterranean seminars*, pp. 341–344.

Ragkos, A., Koutsou, S., & Manousidis, T. (2016b). In search of strategies to face the economic crisis: Evidence from Greek farms. *South European Society and Politics, 21*, 319–337. https://doi.org/10.1080/13608746.2016.1164916.

Ragkos, A., Koutsou, S., Theodoridis, A., Manousidis, T., & Lagka, V. (2018). Labor management strategies in facing the economic crisis. Evidence from Greek livestock farms. *New Meditterranean Journal, 17*(1), 59–72. https://doi.org/10.30682/nm1801f.

Ravis-Giordani, G. (1983). Bergers Corses. In *Les communautés rurales du Niolu*. Marseille: Edisud.

Sandu, D. (2005). Emerging transnational migration from Romanian villages. *Current Sociology, 53*(4), 55–82. https://doi.org/10.1177/0011392105052715.

Simula, G. (2019, February 15). Should we cry over spilled milk? The case of Sardinia. In *Pastoralism, uncertainty and resilience. A blog about learning from pastoralists on how to respond to uncertainty*. https://pastres.wordpress.com/2019/02/15/should-we-cry-over-spilled-milk-the-case-of-sardinia/

Sivini, S. (2019). Azienda agricola "La Capra Felice": l'esperienza di una donna etiope in Trentino. In RRN (Rete Rurale Nazionale). *Terreni d'integrazione* (pp. 48–50). 3(31).

Terrazzoni, L. (2010). Etrangers, Maghrébins et Corses: vers une ethnicisation des rapports sociaux? La construction sociale, historique et politique des relations interethniques en Corse. Ecole doctorale Economie, organisations, société Paris 10 (Nanterre).

Theodoridis, A., & Ragkos, A. (2018). Greece. Unfair trading practices and market outcomes in the Greek dairy sector and the role of hired labor in its development. In M. Migliorini (Ed.), *FOOD TRACK. A transparent and traceable supply chain for the benefit of workers, businesses and consumers: The role of a multisectoral approach to industrial relations and corporate social responsibility* (pp. 13–54). Rome: CGIL.

Vaccaro, I., & Beltran, O. (2007). Consuming space, nature and culture: Patrimonial discussions in the hyper-modern era. *Tourism Geographies, 9*, 254–274. https://doi.org/10.1080/14616680701422715.

Chapter 7
Conclusions

Agriculture and the rural world express the contemporary contradictions of the neoliberal global world; as such, they represent relevant domains to tackle the challenges associated with the intense migratory processes reshaping our societies.

At the crossroads amongst different flows and trajectories, the Mediterranean provides an intriguing setting to analyse the migratory dynamics reconfiguring rural areas. On the European flank there is ample evidence that, on the one hand, agriculture and the rural world hold important potentials for fostering migrants' economic and social integration, as attested by several initiatives. On the other hand, immigrants play a key role in ensuring rural areas remain alive and productive, although these contributions are hardly recognised and appreciated.

In several rural settings, immigrant communities and workers have come to replace a declining local population; immigrant shepherds, for instance, play a key role in ensuring the resilience and persistence of agro-pastoral practices that characterise marginal areas where agricultural intensification proves unfeasible.

The interfaces between agriculture, rural development, and migration are not only fertile in academic terms, but in socio-economic and political ones as well, as the sustainability of these processes requires a comprehensive, integrated policy framework that demands consistency amongst the agricultural, migration, and labour market spheres.

Intense migration is reshaping our societies, raising questions about both the sustainable integration of newcomers in the areas of destination and the impacts in the communities of emigration. Agriculture and the rural world represent relevant settings to tackle these themes as these are increasingly reconfigured by migratory phenomena. At a time when society perceives immigration as a threat to local culture and traditions, evidence from rural contexts conversely shows that immigrants play an important role in maintaining and reproducing local societies and their embedded heritage, including through economic contributions, key social functions, and ecological services.

© The Author(s) 2020
M. Nori, D. Farinella, *Migration, Agriculture and Rural Development*, IMISCOE Research Series, https://doi.org/10.1007/978-3-030-42863-1_7

This volume takes a regional perspective, looking at the Mediterranean. In the region ecological, economic, and socio-demographic asymmetries characterising its different flanks provide relevant push and pull factors for rural migrations. The Mediterranean basin therefore provides an interesting context to assess how migratory flows are reshaping socio-cultural and agro-ecological landscapes and to analyse the differentiated impacts on its different shores.

On the region's eastern and southern rims, limitations in water sources, fertile soils, and reliable climate or peace conditions seriously affect the agricultural livelihoods of a growing rural population; for communities in these areas, emigration represents today an important strategy to expand and diversify the livelihood base.

Conversely on the Mediterranean's northern, European rims, the polarisation of rural development provides important triggers for migratory dynamics. In areas with greater potential for agricultural production, the immigrant workforce proves to be a main pillar underpinning the intensification of most farming systems; the social costs associated with these processes are though high, with increasing concerns over immigrants' living and working conditions.

In more marginal rural settings heavily impacted by the recent economic crisis and public spending cuts, immigrant communities have become a vital asset for local economies and societies. In Europe, mountainous communities, inland areas, and islands provide limited opportunities for income and employment, as well as fewer opportunities for accessing social, cultural, and institutional services. These areas face acute problems of population decline and abandonment, which in turn generate growing concerns related to local generational renewal. These terrains are not only incrementally marginal, but risk being deserted, territories without "societies," with the consequential loss of the local ecological and socio-economic heritage. In such settings immigrant communities increasingly play a critical role in filling the socio-economic gaps left by the national population.

From a European perspective, experience indicates thus that while agriculture and the rural world, on the one hand, can provide important livelihood options to immigrant communities, they are, on the other, themselves a key factor of resilience for many agricultural farms and rural areas.

Agro-pastoralism, the extensive rearing of livestock that characterises most marginal Mediterranean rural settings, provides a useful lens for assessing these dynamics and understanding the contributions and the implications of immigrants on the European rural fabric. Agro-pastoralism embodies the contradictions of globalization since it is increasingly appreciated for the products and services it supplies (from quality animal proteins to biodiversity conservation and landscape maintenance), while also increasingly threatened by global competition and a growing agricultural squeeze that reduce its attractiveness to local populations. The rising shortage of a skilled and committed workforce in this sector is currently buffered by the consistent presence of immigrant shepherds who importantly contribute to keeping agro-pastoral areas alive and productive.

As is the case for most agricultural activities, these contributions are, however, poorly acknowledged, and immigrants operating in rural areas endure heavy exploitation, enjoy limited rights, and remain socially vulnerable. Options for their progress and upscaling are restricted as they face scant prospects for socio-economic improvement.

Contrary to what has historically taken place in the Mediterranean, where population movements have contributed to filling the gaps left by local inhabitants, today this capacity to translate geographical mobility into social is limited: several economic, social, and administrative factors constrain immigrants' evolution from workers to entrepreneurs and farmers in their own right.

The inability to recognize the local relevance of immigrant communities and integrate them into rural development patterns hold relevant implications for the sustainability and the reproduction of the agrarian world in the broader EU setting.

These dynamics raise questions that are both relevant and controversial; the interface between agriculture and migration is fertile not only in academic terms, but in socio-economic and political ones as well. Societal efforts to disentangle and redress these dynamics are needed at different levels, including for researchers, development practitioners, local authorities, policymakers, and civil society alike.

In academic terms, migratory flows in rural areas result in diversified and dynamic socio-cultural fabrics, challenging the idea of a conservative and static agrarian world. Mobilities contribute to the reconfiguration of the rural world as a point of continuous transition, creating new circularity between cities and the countryside, but also transversal networks between regions of different countries.

Most research efforts focus on investigating the drivers and the patterns of migratory flows to and from rural areas, and the diverse implications in the different settings. However, this type of research must be more complex, avoiding functionalism and the over-consideration of economic factors in order to better enhance the understanding of the social, political, and cultural aspects underpinning but also resulting from migratory flows. By better accounting for migrants' subjectivity, recent approaches that look into trans-local mobility and focus on the agency aspects of migrants provide intriguing insights into rural development perspectives.

In policy terms, the reconfiguration of agricultural and rural development patterns is heavily impacting natural and human landscapes throughout the Mediterranean. In Europe concerns are raised on the sustainability of current paradigms informing policymaking, as the longstanding commitment the EU displays through its Common Agricultural Policy does not seem effective in reversing several negative trends.

As a structural factor of agricultural production and rural development in contemporary Europe, the immigrant labour force should attract the attention it deserves from policymakers at various levels. Efforts aimed at redressing development in rural areas should begin by recognizing and capitalising on the important contributions provided by immigrant communities and workers. It remains otherwise difficult to justify EU concern for and support to animal welfare, wildlife status, landscape functioning and consumer safety while policies remain silent and opaque when it comes to the rights and conditions of many agricultural workers and rural dwellers.

Enhancing the integration of immigrant workers in less precarious, longer-term positions and roles in the agrarian world provides a unique opportunity to revitalise depopulating rural areas and support agricultural activities lacking young, skilled, and motivated operators. In such a framework, sustainable agriculture and rural development cannot be merely the result of subsidies, schemes, and incentives, but must be the outcome of a comprehensive integrated policy framework that demands consistency and coherence amongst agricultural, migration, and labour market polices. Adequate decision-making and strategic investments are needed to ensure that rural migrations bring mutual benefits to all stakeholders to reflect the Europe 2020 vision for a "smart, sustainable and inclusive development".

This is obviously not to state that the future of agricultural lands and rural communities stays in their capacities to attract and absorb immigrants, nor that the future of migratory flows resides necessarily in rural areas. Likewise, there are no easy recipes or good practices that apply automatically in supporting local integration of foreign newcomers in rural settings; experiences reported in this volume indicate that these processes are complex and often conflicting, requiring continuous adjustment and mutual adaptation amongst all concerned actors involved.

The risks related to a "ghettoization" effect are obvious, as much as those related to the perception of the rural space as an "empty" and "to be filled," thus ignoring the presence of local communities and helping compound the sense of local marginality. This can produce a double process of "subalternization" in which both migrants and territories are "represented" without the freedom to act and to be. The optimistic idea of countering abandonment and decline and of revitalizing the local economy by "forcibly placing" newcomers in rural areas trivializes both the complexity of the "welcoming" territory and that of those who should "be accepted." It is yet another example of the objectification and abstraction with which these themes are often addressed in public debate.

In fact, the opposition between "migrants" and "natives" is misleading and reductive. Perhaps we need to SIMPLY speak about people living or moving on a rural EDGE where it could be possible to remain provided that concrete opportunities for a dignified livelihood exist and THAT social mobility paths are effective.

Agriculture and the rural realm embody and express the contemporary contradictions of the neoliberal global world. On the one hand, rural areas are the sites of exodus, population decline, economic crisis, and land abandonment or social exploitation. On the other, these represent the space for autonomy, peasant agriculture, multifunctionality, diversity, and resilience, as much as opportunities for integration and cooperation. The new agrarian question needs to develop patterns of inclusive and fair development that account for the needs and the capacities of all actors, including immigrant communities.